Road Vehicle Aerodynamics

ROAD VEHICLE AERODYNAMICS

Second Edition

A. J. Scibor-Rylski, *Ing. Dipl., M.Sc., A.F.A.I.A.A., A.F.R.Ae.S., M.I.Mech.E.*

Second edition revised by

D. M. Sykes, *M.Sc., C.Eng., M.R.Ae.S.*
Department of Aeronautics, The City University, London

A HALSTED PRESS BOOK

JOHN WILEY & SONS
New York

Published in the U.S.A.
by Halsted Press, a Division of
John Wiley & Sons, Inc., New York

First edition 1975
Reprinted 1979
Second edition 1984

Library of Congress Cataloging in Publication Data

Scibor-Rylski, A. J.
Road vehicle aerodynamics.

"A Halsted Press book."
Bibliography: p.
Includes index.
1. Automobiles–Aerodynamics. I. Sykes, D. M.
II. Title.
TL245.S34 1984 629.2'31 84-13196
ISBN 0-470-20097-9

Printed in Great Britain

CONTENTS

PREFACE TO SECOND EDITION

Whilst the material for the first edition of this book was being gathered together by Adam Scibor-Rylski in the early 1970's, the major thrust of research and development work in road vehicle aerodynamics was the result of the efforts of a small group of engineers and scientists. The volume car manufacturers were aware of the science of aerodynamics, and some even encouraged their designers to spend a little time in studying the implications and applications of this science to their art.

In 1974 the drag coefficient of European saloon cars ranged between 0.38 and 0.52, with a simple average value of 0.46. At the same time aerodynamics also had significant impact on the design and appearance of cars used in competitions, with the first appearance of a wing-shaped device on an American Chapparal. The Formula 1 racing cars carried such devices from 1967 onwards and by 1971 the 'add-on' wings were fairly well integrated into the design of the vehicle body. However, in the early 1970's very few people were seriously concerned about the fuel economy of road vehicles. No pressure was apparent to cause car stylists and designers to give more than a passing thought to aerodynamic matters and possibly to find some small applications in areas where there was no conflict with the perceived sales benefit which appeared to follow the artistic expression of the stylists. The commercial transport industry was similarly disinclined to consider fuel economy seriously and no significant attempts were made to reduce the drag of lorries or buses. The plentiful supply of relatively cheap oil and petroleum fuels encouraged such attitudes throughout industrialised society and vehicle operators were no exception.

This changed at the end of 1973 when supplies of oil from the Middle East were threatened and the cost of petroleum based fuels increased dramatically. Petrol and diesel fuel prices increased by 30% almost overnight and the real price of petrol increased by 50% over a twelve month period. This made fuel economy a topic of social and economic importance, and one obvious way of improving the fuel economy of transportation was to reduce the drag coefficient of vehicles at the design stage. The imposition of speed limits achieved some fuel saving over a short period, but improvements in the aerodynamic design of vehicles do not increase operating costs and are effective right through the life of the vehicle.

The first edition of this book appeared in 1975 and it brought together for the first time much information which had been published in various journals and reports dating from the 1920's. The material was put together in such a way that it served many needs and users. The stylist was able to comprehend quickly the aerodynamic and hence operational cost of fashions and trends in body shaping, and he was also guided towards means of achieving desirable objectives. The design engineer was able to better assess the costs and benefits of detailed and overall design changes, and was able to find guidance on good design points. The research engineer and scientist was provided with a state of the art survey which enabled him to quickly identify areas where new ideas were needed and where research and development resources could most effectively be utilised.

At the end of 1983 we find that this book has come to its second edition and the opportunity has been taken to update and extend some of the material which it contained. International Formula 1 racing has seen the incorporation of full 'ground effect' design of body undersurfaces aided by skirts to generate negative lift over the full plan area of the racing car. In five years following the first introduction of this concept by Lotus in 1977, negative lift increased threefold over that available from front and rear aerofoils alone, and exceeded twice the weight of the car. These developments have had their effect on the road holding, braking and cornering characteristics of the racing cars. New regulations will now require that the underbody surfaces of racing cars are plain and have prohibited skirts, but aerodynamic developments will continue.

Many of the lessons contained in the first edition on the interaction between styling and drag have been well learned. Much more attention is being paid to detail design as it is now economical to spend time and effort in such endeavours. Such are the competitive and economic pressures on the manufacturers and designers that whilst in 1976 it was considered that a drag coefficient of 0.30 was out of reach for cars designed under the usual time and cost limitations, such cars were on show in 1981 and on the roads in 1983. Over the past few years, experimental testing has brought increased knowledge about the flow around wheels and wheel cavities, engine cooling requirements and the associated drag penalty, factors affecting the drag and lift of the front and rear end shapes used for popular, mass-produced cars, and other details. Many heavy goods vehicles have had drag reducing devices fitted as add-on items whilst new bodywork designs often incorporate such concepts from the start. All of these different aspects of aerodynamic design are found to interact with performance, fuel consumption and the economics of operating a road vehicle, and relevant information has been introduced into this new edition.

Whilst it is true that the laws of aerodynamics are the same for General Motors, Ford and Volkswagen, it is perhaps most encouraging to observe that the diversity and richness of form and shape is still present in the aesthetic and practical products of the various groups of engineers and stylists engaged in road vehicle design.

D. M. Sykes

PREFACE TO FIRST EDITION

The motor vehicle plays an important part in the life of our society—practically everybody is either an owner or at least a user. A system of fast motorways spanning continents makes it possible to reach and maintain high speeds which together with increasing traffic intensity creates special problems. The interaction between road, atmosphere and car introduces aerodynamic forces which even at comparatively low speeds begin to exceed other forces.

The subject of this book is the analysis of the airflow around a motor vehicle and the relationship between its shape and the resulting aerodynamic characteristics. A discussion of the effects of aerodynamic phenomena on car behaviour is presented in a descriptive and graphical manner, with a minimum of mathematics, which should be comprehensible to the car driver as well as vehicle designers, stylists, aerodynamicists and anyone interested in the design and building of a motor vehicle. The professional stylist-designer will find here an outline of the interrelation between the choice of vehicle shape and the consequent effect on aerodynamic forces and road behaviour. References are given to more advanced aerodynamic textbooks when further specialised investigation is required.

Aerodynamic problems must be given proper attention at the primary stages of car design. The gradual development of the vehicle project from its preliminary stages, through scale model tunnel testing is, therefore, used to illustrate the relationship between car shape and the resulting aerodynamic characteristics.

It is not sufficient, although it is important, to create a low drag shape, since we have also to pay attention to stability, handling and traction problems, all of which are greatly affected by the aerodynamic characteristics of the vehicle.

The effect of aerodynamic forces on performance i.e. speed, fuel economy and steady turning characteristics are, therefore, considered in detail. The book emphasises the influence of aerodynamic forces on road holding in both steady and transient atmospheric wind conditions. In the case of transient atmospheric wind gusts the response of the vehicle, and consequently of the driver, are treated separately and the resultant behaviour of the vehicle discussed. The results of research carried out by the Author on various aspects of vehicle aerodynamics are presented and in reviewing the present state-of-the art the fields in which further research is required have been indicated.

Some anxiety may exist that a strict application of aerodynamics to car styling would tend to make all cars look similar. This raises certain philosophic implications. Aircraft are styled according to the principles of aerodynamics and still maintain their individuality.

Design and styling is the art of compromise. An understanding and application of the laws of nature produces not a monotony but a great diversity and richness of forms and shapes. This truth is easily detected when we observe the world around us.

What is involved is a basic concept of the beauty of a car shape. Vehicle designs which have shapes that contradict natural laws, such as the laws of aerodynamics, would strike us as ugly. Such shapes have something in their appearance which would irritate our sense of proportion. This has of course not only an aesthetic but also a practical and economic aspect. The beautiful car sells well.

It can be argued that the sense of beauty of the general public can be artificially changed by means of consistent propaganda such as mass-media advertising. This method can create a preference for, or rather an acceptance of, certain 'fashions' but it has to be constantly sustained by further expensive doses of artificiality.

In the car business such methods require a constant variety of 'fashion models' to maintain a high level of sales. Thus car styling is sometimes affected by 'market research' investigations which claim that the reaction of the public to past creations indicates future preferences.

This work is dedicated to broadening the fundamental basis of car styling philosophy. To achieve this, a knowledge of the fundamental laws of aerodynamics should become an essential part of every car stylist or designers intellectual equipment. In trying to disseminate this knowledge this book sets out to enlarge the aesthetic and practical comprehension of car style and beauty.

A. J. Scibor-Rylski

Chapter 1.

INTRODUCTION

By definition *aerodynamics* is that branch of the science of physics which is concerned with the study of the interaction between a body and the atmosphere through which it moves. The forces which are created by this interaction act upon the moving body and are called aerodynamic forces. The magnitude and direction of these forces depend on the relative velocity of the air and the body, in this case the automobile, and its shape.

Aerodynamic forces created by the motion of a vehicle affect its performance. It is the purpose of this book to describe the aerodynamic phenomena involved and to indicate their importance and influence on practical driving. The professional stylist-designer will find here an outline of the interrelation between the choice of vehicle shape and the consequent effect of the aerodynamic forces on the behaviour of the automobile on the road.

The component of the resultant aerodynamic force which opposes the forward motion is called the *aerodynamic drag*. It greatly affects the performance of the vehicle in both speed and fuel economy as it is the engine power which has to overcome this opposing force. The other component of the resultant aerodynamic force, directed vertically, is called the *aerodynamic lift*. It reduces the frictional forces between the tyres and the road surface, thus changing radically the steering and propelling characteristics of the vehicle. The aerodynamic force is the net result of all the variably distributed pressures which the airstream exerts on the car surface. This 'pressure distribution' varies with the attitude of the automobile relative to the airstream and creates conditions which can be either helpful or detrimental to steering and handling characteristics. The aerodynamic force is, therefore, an important consideration in car stability. This in turn concerns the problem of safety which affects every car and thus road traffic generally.

This approach is reflected in the order of the sections of this book. The first part deals with the analysis of the aerodynamic force and its drag and lift components. The second part deals with problems of the aerodynamic effect on car stability and road holding.

The first stages in the development of self-propelled road vehicles show that they inherited their shape from horse-driven carriages. Designers of the early cars had first to solve the mechanical problems since the design was totally dependent on solutions of a structural kind. In view of the very restricted speed of these vehicles problems of an aerodynamic nature did not exist.

The gradual increase of automobile speed and rapid developments in their design soon meant that drivers and occupants were exposed to the airstream. This in turn led to the introduction of such structural parts as windscreens and gradually the shape of the developing vehicles began to be influenced from the point of view of protecting the occupants from the airstream effect. The concept of streamlining was introduced later when the lighter-than-air airship appeared on the scene.

Streamlining was then directly related to the progress of aeronautics. This process started about 1920 when some early experiments were carried out in Germany which showed the possibility of reducing the aerodynamic drag force.

Systematic research into automobile aerodynamics was organised in Germany approximately 25 years after the first appearance of the automobile. In 1922 one of the first articles on this subject appeared in the German technical press [1]. These articles represented the results of a series of experiments on the effect of body shape on the aerodynamic drag force, carried out in the great wind tunnel in the Zeppelin works in Friedrichshafen.

Research into aerodynamic effects on the stability of cars and on their road holding characteristics was started soon after by a group of scientific workers led by Kamm [2]. As the result of this work, or rather as its logical conclusion, the Research Institute for Vehicular Science was organised in Stuttgart. This institute was reconstructed after the second World War and is at present a part of Stuttgart's University of Technology. Results of their research have influenced other research programmes [3,4,5,6,7,8,9,10]. Interest in car aerodynamics in the U.S.A. began about 1930 (Chrysler) but was not then continued systematically. The post-war American research in this field dates from about 1950 and was carried out mainly by Ford.

Following the first successful car design, further development was mainly involved with mechanical problems concerned with the engine, transmission, suspension and steering. After approximately 25 years of steady development, automobiles reached a stage at which aerodynamic problems again began to be recognised. This coincided with the stage when advances in streamlining of aeroplanes was producing a dramatic improvement in their performance. It is not surprising therefore that aerodynamicists began first by considering the aerodynamic drag problem of road vehicles. Other connected problems such as aerodynamic lift and its effect on car stability only attracted attention later when vehicle speed had increased to such an extent that its effects could no longer be ignored.

To illustrate the complex interrelation between the development of vehicle shape and its aerodynamic characteristics it is useful to study the gradual evolution of mechanical design and its influence on external shape.

In about 1956, in the United States, due partly to the influence of a few specialists and partly to the logical and systematic approach to such problems in the U.S.A., vehicle shape was studied analytically from the point of view of its functional development [11]. This analysis was carried out on the assumption that an object which moves in a given medium tends to assume a form which offers the least resistance to motion. Such a development affects the vehicle functionally, moulding its external shape. The natural evolution of the vehicle is thus a measure of progress. The more complete the adaptation of the form of a vehicle to its environment the more advanced and the better the technological design. By the term 'adaptation' we mean internal mechanical perfection integrated into the external shape. Such morphological changes may be considered as a result of the interaction between the vehicle and its environment.

The process of development has its own history in which rational design trends mix with some accidental occurrences. Consider, for example, the most fundamental dimensions of a vehicle presented as a rectangular prism (Fig. 1.1). The relation between the height to length ratio C/A and the width to length ratio B/A, Fig. 1.1(a), shows that passenger cars have a ratio $B/A > C/A$, while for commercial vehicles $B/A < C/A$. This may be explained, in the case of commercial vehicles, by the natural tendency to accommodate bulk cargo on a given chassis and in the case of saloon cars by the development of a body shape wrapped around passengers for their comfort.

Road conditions have limited the width of vehicles. It is said that this width is the inheritance from the times of horse-drawn carriages when the width was established naturally by two horses running comfortably side by side. Automobile length is an indirect function of the width and, generally, is less restricted. Thus length is more independent of the influence of the environment. In general vehicle weight increases with length. Passenger cars under two tons have reached their maximum length since restrictions on further increases are limited by road clearance, an acceptable road turning radius and the requirement of manoeuvrability, Fig. 1.1(b). Therefore, morphological developments* throughout the years have produced certain limitations on dimensions around which the external shape is wrapped.

A historical survey of the variation in the aerodynamic drag component, which was the subject of comparatively early investigation,

* By morphological development we mean the evolutionary change influenced by the environment.

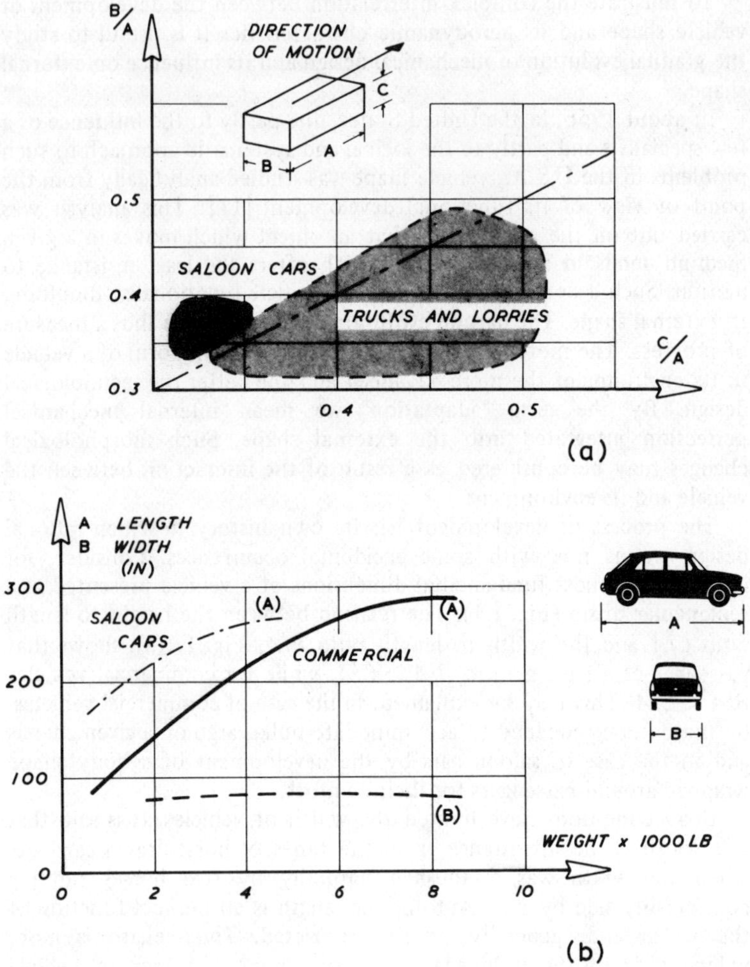

Fig. 1.1. (a) The main dimensions of road vehicles, (b) relation between car weight and general dimensions

serves as one more example of changes in car design and shape due to the influence of road conditions and usage. In 1904, when wheels were, in general, still separated from the main body of the vehicle, the frontal windshield was usually almost vertical and the bucket seats of the driver and passenger were exposed to the airflow; consequently the aerodynamic drag was very high. The addition of wheel mudguards and later the joining of them to the main body of the vehicle, which at a later stage led to the development of the present saloon car shape, is indicated in Fig. 1.2(a) by the gradual reduction of the drag. The vertical scale in Fig. 1.2 represents fractional values of the initial high drag value which is taken as unity. This trend continued with further modifications, such as the merging of the headlamps with the main body, and by about 1950 the average aerodynamic drag had reached its minimum value of approximately 45% of the initial aerodynamic drag of cars of 1904 vintage. After 1950 this trend was reversed and the drag started to increase. Close investigation of this phenomenon seems to indicate that this increase is related to the filling-up of the space between the bonnet and the mudguards and consequent change of the planform of vehicles from triangular to rectangular. As further investigations discussed in this book will show these styling trends caused, in effect, a change in the general flow pattern inducing the airflow to divide more sharply into flow over and under the vehicle instead of, as before, inducing it to divide sideways. Thus the rectangular planform of more modern vehicles, in comparison with the more triangular shape of older vehicles, seems to create greater aerodynamic drag. Thus the development in the design and shape conforms to the natural laws of nature.

Other reasons have affected the morphological development of the 'cross-country' vehicle whose shape is more susceptible to the influence of the environment–the reason for this lies in the terrain over which these vehicles move and the type of obstacles which they have to overcome. The development of the sports car and of the saloon car have surpassed that of the cross-country vehicle and this is mainly due to the more advanced, and in a way simpler, adaptation of their forms to road conditions. The racing car can be considered as the ultimate, although somewhat atypical, example of such a development.

On the other hand, as we mentioned earlier, external shapes are dependent on mechanical design, such as engine size and the design of transmission and suspension mechanisms. Other advanced technologies like aeronautics have shown that correct integration of mechanical and other systems into a complex well-designed machine is essential for its advance and development. Therefore the design of the constituent parts and their integration into a complete design plays a very important role in the morphological development of the automobile. We conclude then that a close affinity exists between design, the integration of the constituent parts and the body shape, which in turn affects their performance in the road-atmosphere environment in which they move.

The modern road system, and consequently the general progress of methods in transportation, emphasises the factor of speed in the morphological process. The gradual increase in the average speed of cars

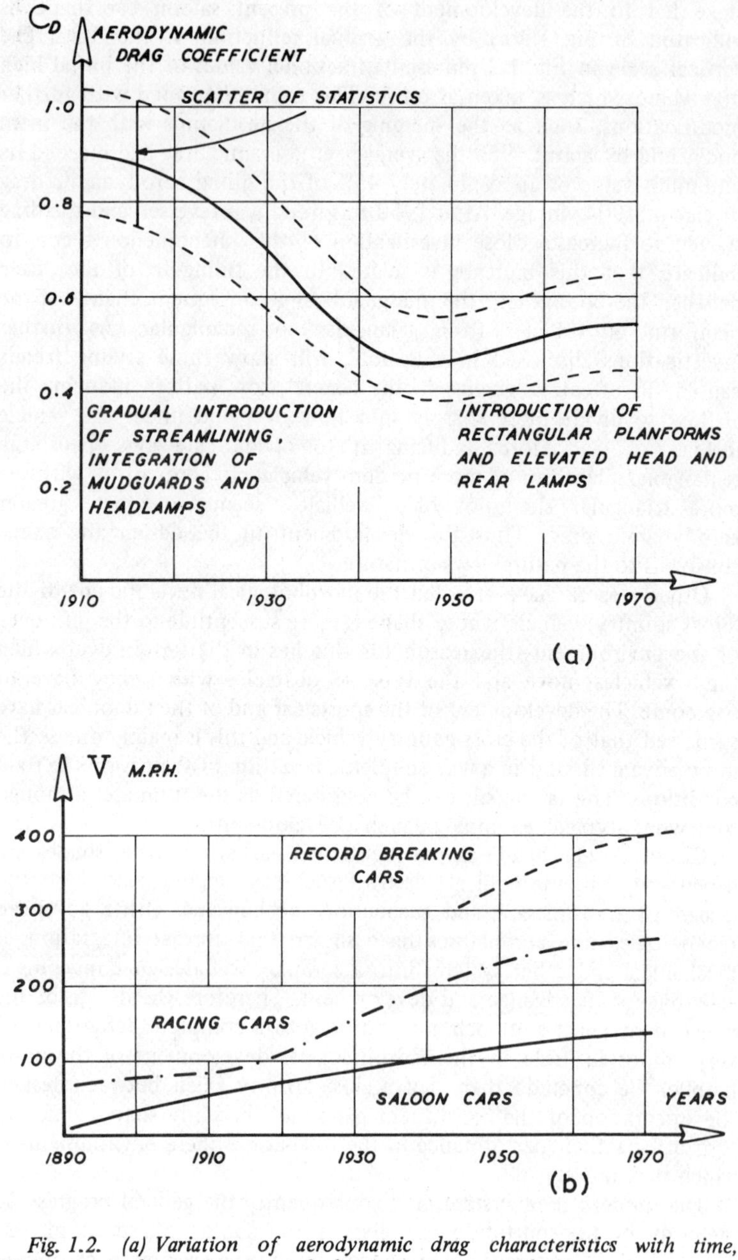

Fig. 1.2. (a) Variation of aerodynamic drag characteristics with time, (b) increase in the maximum speed of various car types with time

is shown in Fig. 1.2(b). The motorway systems consisting of broad multi-lane roads with wide and openly inclined sides are much more affected by atmospheric winds than the former narrow winding roads which they replaced. This causes cars to be exposed more often to winds and wind gusts which enhances the importance and underlines the increasing degree of preoccupation with aerodynamic effects.

The laws of aerodynamics are well known. These laws are applicable in all cases where there is a relative velocity between the air and a vehicle. For a long time the application of aerodynamics was practically exclusively aeronautical, and it is to aeronautics that we are indebted for the great progress in this field. Road vehicle designers and stylists have benefited from this great accumulation of experience. Further and progressively deeper study of car aerodynamics, both in theory and in the experimental field, shows clearly some fundamental differences between aeronautics and automobile application.

The aeroplane moves in free flight while the motion of a car is restricted by the ground on which it moves. Therefore the disturbance to the atmosphere caused by an aeroplane in flight is freely three-dimensional while the airflow around a car is distorted by the interference of the ground.

The car in motion is subjected to inertia and frictional forces to which the externally applied aerodynamic forces are additive; here the frictional forces created by the tyres on the road are of primary importance. It is therefore now increasingly recognised that car aerodynamics has to be considered quite independently of aeronautics, and aerodynamics specific to cars must be developed.

Aerodynamic forces act externally on a car in combination with other forces of a mechanical nature, like traction-propellent force, tyre-road frictional reaction force, and mass-inertia force. It would therefore be unreasonable to consider aerodynamics separately from the mechanical aspects.

The importance of aerodynamics is nevertheless such that it is difficult, even impossible, to predict the performance and general road behaviour exactly without having first established the aerodynamic characteristics. Problems of aerodynamics should be solved before the actual construction of the car prototype, and then checked on a full-sized vehicle. The stylist-designer should be familiar with the basic aerodynamic principles involved when undertaking the preliminary stages of design and deciding the car shape. The preliminary design project, when it is at the stage at which the external shape is being established, should allow for the vehicle's aerodynamic characteristics to be estimated at the earliest moment as it will enable its probable performance, stability and road holding characteristics to be calculated.

It is cheaper to investigate these problems on scale models in a wind tunnel. This technique is well known and highly developed in aeronautics. The preparation of a scale model requires from the start a sound general knowledge of aerodynamics, and especially car aerodynamics. When styling proposals for a new project are under investigation a clay model (usually 1/5th scale) is used to allow for changes in detail to be made during the progressive aerodynamic testing. German designers call it colloquially 'aerodynamic tuning'. Some typical aerodynamic characteristics such as drag, lift, side force and aerodynamic yawing moments could be estimated at an early stage by various empirical but crude methods. These methods are based on the comparison of the suggested vehicle shape with the geometrical likeness of known vehicle shapes [12,13] to enable average values of the aerodynamic forces to be calculated. These 'first approximation' methods can be applied when the proposed car is similar to a known shape.

In problems of stability, which also should be considered at a very early stage of design, the value of the aerodynamic force is not as important as the rate at which it varies with the yaw, pitch or roll angles of the vehicle. This is more difficult to assess without proper experimental investigation; approximate calculations often fail to produce reliable values and cannot be a substitute for aerodynamic investigation.

Our accumulated experience and know-how of car aerodynamics is gathered from many scale models tested in wind tunnels and from investigations on full-sized vehicles carrying measuring equipment which are tested on the road under natural conditions.

The reader will find in the following chapters a description and analysis of the aerodynamic problems of cars from the point of view of both the user and designer.

Chapter 2

AERODYNAMIC FORCES AND MOMENTS

2.1 GENERAL AIRFLOW PATTERN AROUND A CAR

When an automobile moves along a road it moves relative to the ground. When the air is still and there is no wind the car has the same velocity with respect to the ground as it has with respect to the air. But when there is an atmospheric wind and the air moves relative to the ground the car has a different velocity with respect to the ground than with respect to the air.

To simplify our considerations of car aerodynamics we shall assume initially that the atmosphere is at rest with respect to the ground, i.e. there is no wind. Also, we shall assume that the speed at which the car is travelling is constant and, throughout this chapter, that the vehicle is moving along a road which is level.

It is important to realize that in all cases the aerodynamic phenomena are the same in principle, whether we consider air moving past the car (as simulated in wind tunnels) or, as in a real case, when the car is moving relative to the air. All aerodynamic phenomena are due to this relative motion of the air over the car shape.

Consider a particle of air passing through *A,* in Fig. 2.1(a), at a given moment in time. It moves relative to the axes *xyz,* fixed to the moving car, and follows a path which we call a *streamline.* As we assumed a constant speed for the car the next air particle finding itself at point *A* would follow exactly the same path, and so on for all other particles passing through *A.* Therefore we can consider the streamline as fixed in space and belonging to a family of streamlines which form an *airflow pattern* around the car. This airflow pattern depends on the car shape and on the speed of the car. Such an airflow pattern can be visualised by different methods using either a full sized car or a model in a wind tunnel. One of the more frequently used and practical methods is fixing flexible tufts on to the surface of the car. The local airstream orientate these tufts in the direction of the local flow. Thus the general pattern of tufts indicates the airflow pattern. In some cases it is better to film it as this would show the variation with time, but when the airflow is fairly steady, a still photograph gives a good idea of the local flow pattern. Fig. 2.2 shows a racing car with such tufts during a test run.

The streamlines far ahead of the car will be parallel and undisturbed while in the vicinity of the car they assume a more complex pattern.

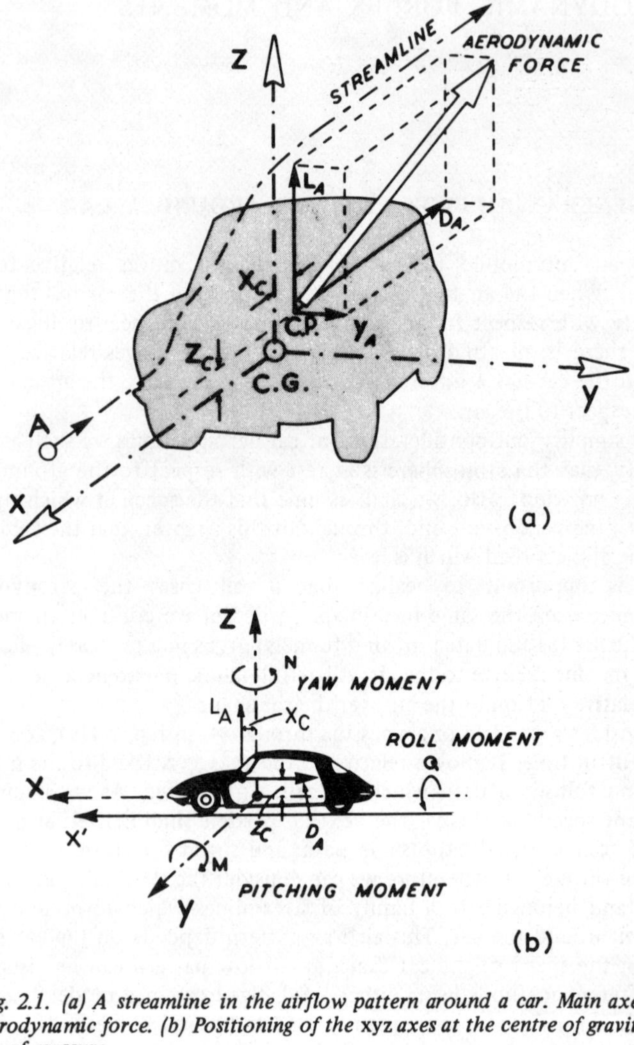

Fig. 2.1. (a) A streamline in the airflow pattern around a car. Main axes and the aerodynamic force. (b) Positioning of the xyz axes at the centre of gravity and centre of pressure

Thus around the car we have a region of flow disturbance. Studying the particle motion we find that far ahead of the car the relative velocity of the particles are equal to the car speed, while within the disturbance region around the car the relative velocities of particles vary considerably, being either greater or slower than the actual car speed.

Fig. 2.2. *Flow visualisation method (tufting) used on a racing car. (By courtesy of Ian Bracey, Clubman Formula IBEC PI racing car)*

2.2 ORIGIN OF THE AERODYNAMIC FORCE

From the general law of the conservation of energy we can derive a well known aerodynamic formula. This formula expresses the fact that the sum of the atmospheric (ambient) and dynamic pressure is constant at every point of the same streamline. Denoting the atmospheric pressure as p, the dynamic pressure as $q = \rho V^2/2$ and the total pressure as H, we may write this fundamental aerodynamic law as follows:

$$p + \frac{\rho V^2}{2} = H \text{ (constant)} \tag{2.1}$$

This is known as Bernoulli's theorem. The value of the constant H may be calculated from airflow conditions far ahead of the vehicle. This formula shows that when the local airstream velocity V varies (and hence the local dynamic pressure) the value of the local atmospheric pressure must vary also, so that the sum of the pressures remain constant. Therefore, where the velocity of the local airstream over the surface is greater than the velocity of the undisturbed stream, the value of the dynamic pressure q is higher, and consequently, the value of the local atmospheric pressure is lower than that of the undisturbed stream, and vice versa. A practical application of this law indicates that the airflow pattern around a car produces a specific pressure distribution over its surface.

Integrating these pressures over the total external car surface we obtain the resultant force on the car, exerted by the relative motion of the air. This we call the aerodynamic force, F. Although this force is the resultant of all the pressures acting on the external surface it is usually, for convenience, replaced by a single resultant force applied at a specific point called the *centre of pressure,* denoted in abbreviated form as the C.P.

For a further description of fundamental aerodynamic principles the reader is referred to the textbooks listed in the bibliography [14,15,16].

It is found that the aerodynamic force is proportional to the square of the air velocity, to the area exposed to the airstream and to a non-dimensional coefficient dependent on the shape of the body, and can be expressed as follows

$$F = qSC_F \tag{2.2}$$

Where S is the reference area and C_F is a non-dimensional coefficient. The resultant aerodynamic force (Fig. 2.1(a)) is resolved for convenience of the analysis into the following three components:

(1) The aerodynamic drag force D_A which is horizontal and opposes the forward motion of the car.

(2) The aerodynamic lift force L_A which is directed vertically and at right angles to the drag.
(3) When the airflow is not symmetric to the car shape we have a sideways component of the aerodynamic force, Y_A, which acts horizontally but at right angles to both the drag and the lift.

These components of the aerodynamic force can be calculated from Equation 2.2.

$$D_A = \frac{\rho V^2}{2} S C_D$$

which gives the drag coefficient

$$C_D = \frac{D_A}{\frac{\rho V^2}{2} S} \tag{2.3(a)}$$

$$L_A = \frac{\rho V^2}{2} S C_L$$

which gives the lift coefficient

$$C_L = \frac{L_A}{\frac{\rho V^2}{2} S} \tag{2.3(b)}$$

$$Y_A = \frac{\rho V^2}{2} S C_Y,$$

which gives the side force coefficient

$$C_Y = \frac{Y_A}{\frac{\rho V^2}{2} S} \tag{2.3(c)}$$

Aerodynamic moments.

We can transfer these three components of the aerodynamic force to any point other than the centre of pressure providing that we add to them the aerodynamic moments created by the product of these forces and the distance of the given point from the centre of pressure. This is of practical importance because we have to consider the inertia forces acting on the car mass, the frictional forces acting on the tyres and the aerodynamic force components all acting simultaneously. We have therefore to transmit the action of all these forces to the centre of gravity of the car where the mass-inertia forces operate. Thus the aerodynamic moments given are usually calculated with respect to the centre of gravity, see Fig. 2.1(b).

The aerodynamic pitching moment M_A is caused by the combined effect of the aerodynamic lift and drag. Considering the position of the C.P. relative to the C.G., described by distances x_c and z_c we may write

$$M_A = L_A x_c + D_A z_c = qS(C_L x_c + C_D z_c) = qSlC_M \quad (2.4)$$

The symbol l in this equation denotes the *characteristic length* which for cars is usually accepted to be the wheelbase. C_M is a non-dimensional pitching moment coefficient.

The aerodynamic yaw moment N_A is caused by the side force Y_A acting at a distance x_c. Therefore we may write

$$N_A = Y_A x_c = qSl\, C_N \quad (2.5)$$

In this equation the symbol C_N denotes the non-dimensional yaw moment coefficient.

The aerodynamic rolling moment coefficient Q_A is caused by the side force Y_A acting at a distance z_c. Therefore, we may write

$$Q_A = Y_A z_c = qSlC_q \quad (2.6)$$

In this equation the symbol C_q denotes the non-dimensional rolling moment coefficient. It is accepted in the car industry that all such coefficients are related to the maximum cross-sectional area. This area is practically the same as the frontal area and it is often referred to as the 'reference area', denoted as *S.* The coefficients of the aerodynamic moments are in addition calculated in relation to the characteristic length.

The reference area *S* should also include the cross-sectional area of those parts of the wheel which are directly exposed to the airstream. The area *S* is sometimes called the 'frontal area' of the vehicle because it is, in fact, approximately the area visible to an eye, or a photographic camera at a distance sufficiently great (theoretically at infinity) to be able to neglect the spheric angle effects.

A number of authors follow aeronautical engineering practice and define their axes system with Ox forward, Oy going to the right and Oz running vertically downward. Since these axes are fixed relative to the body geometry, a number of workers refer to the fore-aft force as the *X* force and reserve the title 'drag' for the force acting in the direction of the relative wind velocity, which is the force most often measured in wind tunnel experiments.

The following chapters discuss the general airflow pattern around a car.

Chapter 3

AIR VISCOSITY PHENOMENA

3.1 AIRFLOW SEPARATION

Text books on aerodynamics consider air, theoretically, as a gas which does not possess any internal friction, meaning that it is not viscous. Any shape moving in such an ideal atmosphere would have the sum of the pressure integrated over the external surface equal to zero. This fact is often illustrated by constructing a flow pattern of an 'ideal gas' around a cylindrical form. Fig. 3.1(a) shows the symmetry of pressures both in the horizontal and the vertical planes. Such a pressure distribution, when integrated around the circumference of the cylinder (per unit span) would produce a zero aerodynamic force. This result is known in aerodynamics as d'Alembert's paradox.

In reality experiment shows that there is always a resultant aerodynamic force acting on a body in motion. This force is in fact caused by the phenomenon of air viscosity. The atmosphere does not behave like an ideal gas. It is viscous, which means that during its motion relative to a surface it displays internal friction. Particles of air when in contact with the surface of a body are arrested by it and the relative velocity at the surface becomes zero. The layers of air nearest to the surface are then slowed down by the viscous friction and their velocity relative to the surface varies with the distance from it. As the distance increases the particles of air are progressively less affected by the viscosity, they move relatively faster until at a certain distance from the surface they move almost freely, practically unaffected by the viscosity. The way in which a moving body in a gaseous medium imparts its momentum to particles of the gas is through its viscosity. Fig. 3.1(b) shows how the flow pattern around a body changes due to viscosity and consequently how the pressure distribution becomes unsymmetric. Integration of the pressure around the circumference will produce an aerodynamic force which in this case would be simply the drag of the cylinder.

The boundary layer

This viscous effect causes the air layer immediately surrounding the moving body to move with it. Layers at greater distances, but still near, would move with varying relative speeds. Therefore, within a compara-

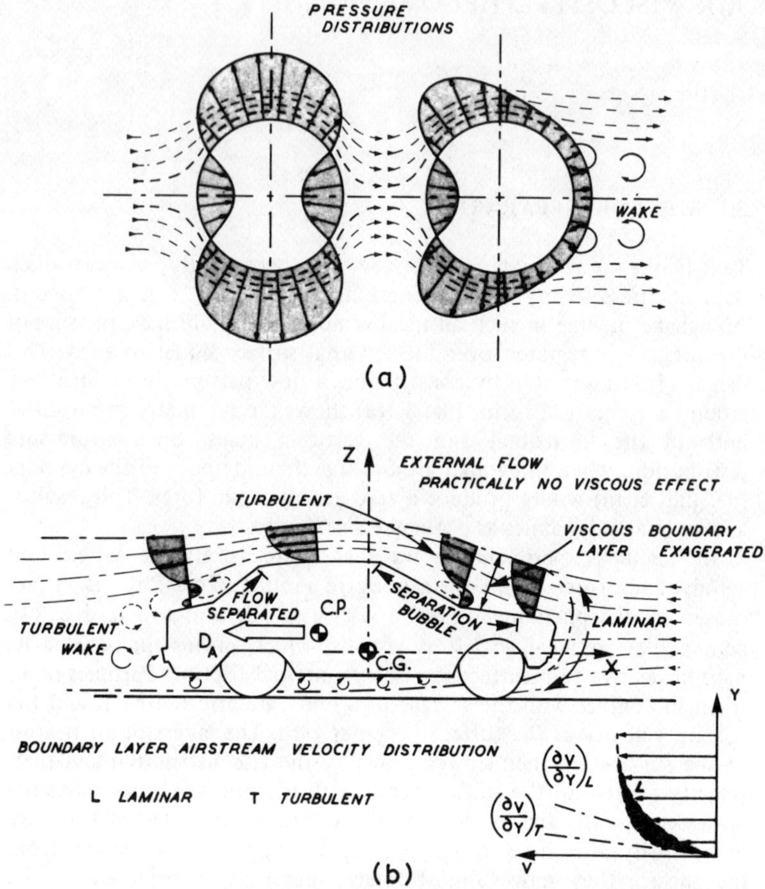

Fig. 3.1. (a) The flow pattern and pressure distribution around a cylinder in an 'ideal gas' or air flow. (b) The boundary layer along the central plane of the upper surface of a car

tively thin layer of air surrounding the moving body air there would be a sharp variation of velocity, from zero at the surface to the local airflow velocity outside this layer. Within this layer there is therefore a *velocity gradient.* The layer itself is referred to as the *boundary layer.* Expressing velocity increase within such a layer in mathematical terms we write the gradient as the derivative $\partial V/\partial y$ where the symbol y denotes the distance from the surface in the direction of the v axis. The manner in which the velocity gradient varies within the boundary layer gives us a clue to the type of the boundary layer. When the gradient is large, i.e. when the rate of velocity change within the boundary layer is great, the friction between one layer and the next is considerable and the whole layer becomes turbulent, i.e. full of small eddies. When the friction inside the boundary layer is not excessive, i.e. when the flow velocity in general is not very great, sub-layers within the boundary layer slide one over the other with different velocities. We call such a boundary layer *laminar.* Thus we have two distinct types of boundary layer, laminar at small local velocities and *turbulent* at higher local velocities. As the airflow pattern involves a wide range of local air velocities we may conclude that both types of boundary layer occur on a car surface. Fig. 3.1(b) shows that only a very small part of the boundary layer in front of the car is laminar the rest is turbulent. We may assume that practically all the car surface is covered by a turbulent boundary layer.

Airflow at the surface causes frictional aerodynamic drag. This is due to the shear stress, or in other words the skin friction. This shear stress, due to the viscosity, produces a tangential force. Because it is created by the viscosity it is directly related to the velocity gradient at the surface.

$$\text{Skin friction } \tau_0 = \left(\frac{\partial V}{\partial y}\right)_{y=0} \mu \qquad (3.1)$$

where μ is the air viscosity coefficient.* The rate of variation of air velocity with distance from the surface characterises the given type of boundary layer.

In a laminar boundary layer the velocity near the surface increases (from zero at the surface) less rapidly than in the case of a turbulent boundary layer. Therefore the value of the derivative $\partial V/\partial y_{y=0}$ is greater in the case of the turbulent layer, i.e. the skin friction involved is greater than in the case of the laminar boundary layer. It would be beneficial for cars to be surrounded by a laminar layer but this is impossible to achieve in practice as the car shape and atmospheric conditions in proximity to the ground are factors which add to induce turbulence.

The effect of air friction and the presence of the boundary layer influence the aerodynamic force in a rather indirect way, i.e. by introducing an asymmetry in the airflow pattern, which includes the phenomenon of 'airflow separation', thus radically changing the pressure distribution.

The growth of the viscous boundary layer on the surface of the car depends on many factors including the local air velocity induced by the

* The viscosity coefficient μ under normal standard conditions, i.e. 1 atmosphere and 15° C is $\mu = 1.75 \times 10^{-6}$ $(T/T_O)^{0\cdot 76}$ kg s/m² where $T_O = 288°$ K and T is the local absolute atmospheric temperature in degrees kelvin.

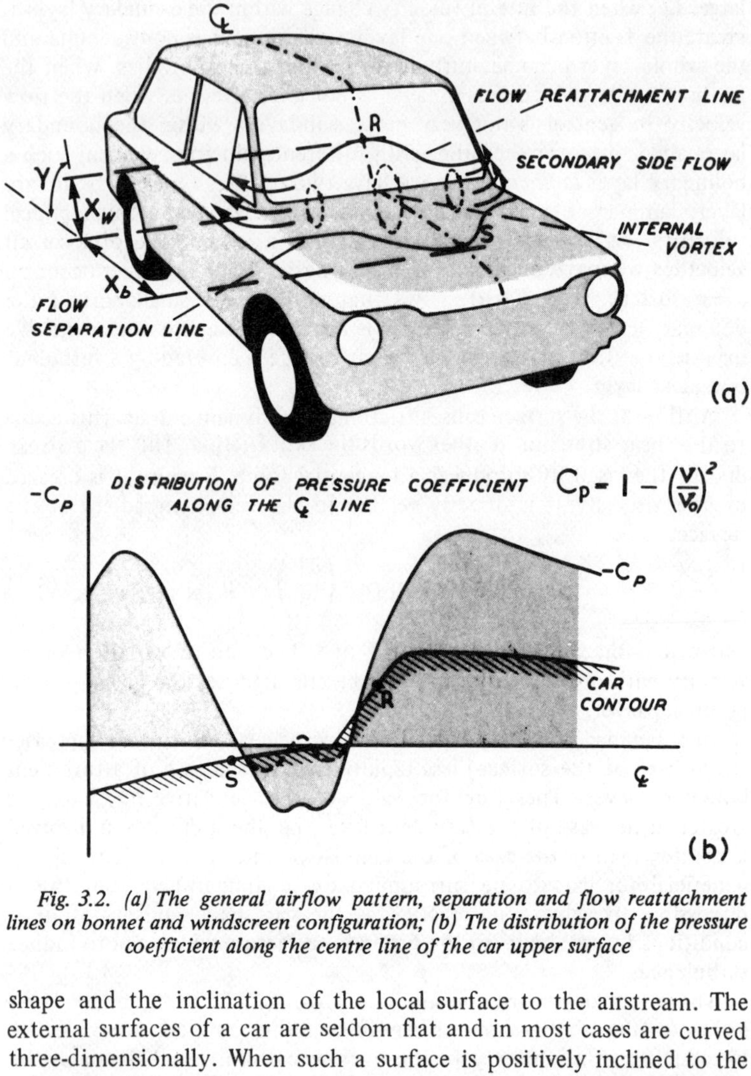

Fig. 3.2. (a) The general airflow pattern, separation and flow reattachment lines on bonnet and windscreen configuration; (b) The distribution of the pressure coefficient along the centre line of the car upper surface

shape and the inclination of the local surface to the airstream. The external surfaces of a car are seldom flat and in most cases are curved three-dimensionally. When such a surface is positively inclined to the oncoming stream, the external flow pattern causes the air to accelerate and, according to Bernoulli's theorem, this causes a decrease in the local pressure. In such regions of the surface, the pressure gradient is negative along a stream line and air is accelerated maintaining a certain boundary layer thickness at the surface.

When the surface is curving away from the airstream the effect is reversed. According to Equation 2.1 the ambient pressure gradually increases due to the slowing down of the airstream. This in turn creates a condition in which the airflow moves against the increasing pressure. We say that the 'pressure gradient' in the flow field is adverse, i.e. $dp/dx > 0$. This affects not only the airstream outside but also, and mainly, the airflow inside the boundary layer. The flow within the boundary layer begins to slow down causing the boundary layer to 'pile up' and thicken. There comes a point at which the particles of air nearest to the surface lose their momentum and at this point friction and the adverse pressure gradient bring the innermost boundary layer particles to a standstill, while the particles further from the skin move downstream, although more slowly. At this point the boundary layer separates from the surface and induces the airstream, generally, to break away. In experimental aerodynamics the point S is called the *separation point,* Fig. 3.2(a).

The body shape can be roughly subdivided into two distinct parts by a plane normal to the general direction of motion, which in practise usually coincides with the maximum cross-sectional area. At the front of the car the airstream is, in general, inclined upwards causing the air particles to accelerate. Any separation of the flow on this part of the surface could be only brought about by artificial means, such as sharp corners. The cross-sectional area of the car gradually reduces towards the back. This in general induces the flow to slow down and creates the conditions described above, facilitating flow separation. As a car body is of limited length the flow has to separate somewhere and it is better to induce separation to occur where it would be most advantageous. The actual points of flow separation form a separation line which is in fact the boundary of the rear wake behind the vehicle.

The wake behind the car may be defined as the volume of air to which the car imparts some of its momentum by means of viscosity. In actual fact this volume of air tends to travel with the car.

3.2 LOCAL AIRFLOW SEPARATION BETWEEN BONNET AND WINDSCREEN

The flow separation is often made visible by dust and dirt deposits on the surface of a car when driven in bad weather. The dirt deposit rate is considerable over those parts of the surface where turbulence predominates, while over parts subjected to more streamlined flow, the dirt is rather 'pushed along' by the air flow, sometimes leaving marked trajectories. Thus the flow pattern is often visible after a long run. It is then visible on the front part of the car, on the bonnet and in front of the windscreen. It is exceptional for flow separation to occur on the frontal part of the car body where it is usually caused by the corner created by the bonnet-windscreen configuration. Although the flow over the front of the bonnet is very fast the presence of the windscreen, especially when the angle with the bonnet is large, results in a slowing down of the airstream causing it to separate in front of the windscreen. Thus a 'separation bubble' is formed at the apex of the corner, i.e. at the bottom of the windscreen. The flow reattaches itself to the upper parts of the windscreen, see point *R* Fig. 3.2. The technological solution of the windscreen-bonnet configuration requires special study. As an introduction to such a study some experimental investigations were made by the Author with the help of final year students of the City University [17].

In the first instance the simple case of two flat plates forming a corner simulating the windscreen-bonnet configuration was investigated in the wind tunnel. Then the model was redesigned to facilitate the investigation of the effect of the windscreen curvature on the extent and positioning of the separation bubble. Some preliminary experimental investigation results are shown in Fig. 3.3. This is a simple case of two flat surfaces inclined to each other at an angle γ. The separation lines and the re-attachment lines on the windscreen are clearly visible. Fig. 3.3 shows the secondary flow from the separation bubble towards the right and left corners. This is also noted in actual cars when the separation is made visible by the dirt deposit. The visualisation of the flow within the separation bubble indicates that a

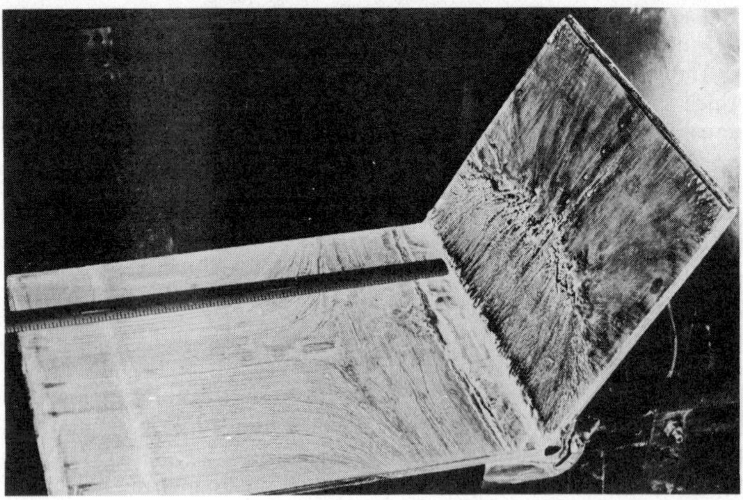

Fig. 3.3. Visualisation of the flow over a corner simulating the simplified bonnet and windscreen of a car

rotary flow, or vortex, exists which extends sideways and probably continues along the sides of the car. This feature of the airflow in front of the windscreen is discussed more fully in a later chapter concerned with aerodynamic lift problems.

The investigation of this configuration has shown that the following main variables influence the airflow over the bonnet-windscreen corner:

(1) The angle γ between the bonnet and the windscreen.

(2) The three-dimensional curvature and configuration of the bonnet.

(3) The three-dimensional curvature and configuration of the windscreen.

The effect of these variables can be investigated separately; (1) is shown in Fig. 3.4 which gives the relative positions of the main separation point (on the line of symmetry) and the point of reattachment R which is plotted with respect to the angle γ. The relation between the positions of points S and R and the angle γ form a circle when plotted [17]. Therefore, values of the non-dimensional distances x/d and x/c can be given in the form of an equation of a circle. The maximum deviation from these formulae for the point R is at $\gamma = 60°$, when the error is of the order of 2% of the overall length of the

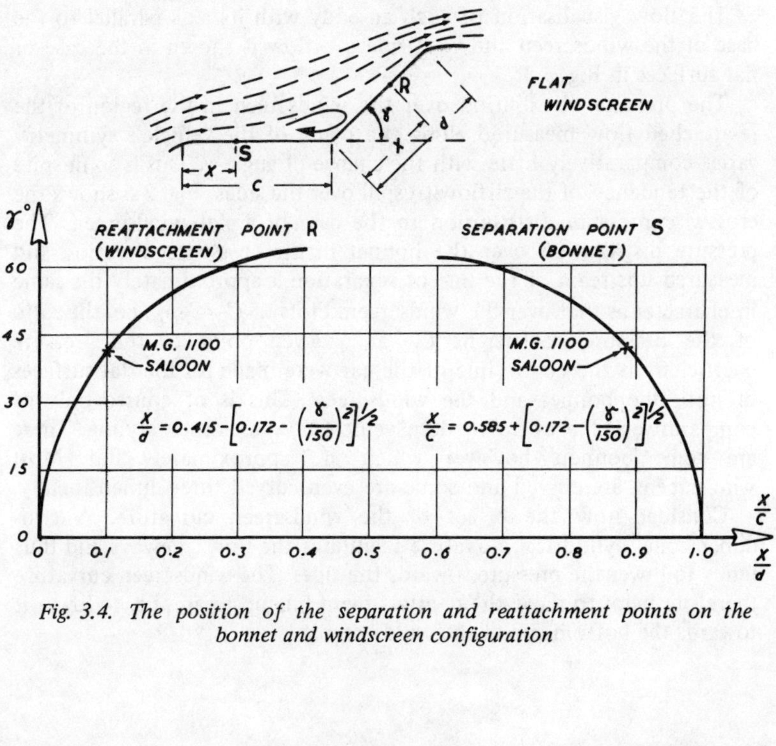

Fig. 3.4. The position of the separation and reattachment points on the bonnet and windscreen configuration

windscreen. The increased curvature of the bonnet, measured horizontally, tends to move the separation point S downstream. Similarly the increase of the curvature of the windscreen tends to move the re-attachment point R downstream.

Experimental research indicates that a small bonnet curvature measured in the horizontal plane does not significantly affect the movement of the point S. The usual current curvatures of the bonnet are rather small ($1/R = 0.02\ \text{m}^{-1}$). Much more influential in this respect is the flow restriction on the sides of the bonnet which can result, for example, from elevated fairings of headlamps. Any restriction imposed on the flow, preventing it from turning sideways, would naturally affect the shape of the separation line and would tend to move it upstream.

The most remarkable result obtained from these experimental investigations [17] is that at a certain angle γ the pressure in the region of the bottom of the windscreen is lowered. The kink in the pressure distribution in this region as measured along the plane of symmetry is shown in Fig. 3.2(b). This local pressure drop occurs when $\gamma = 48°$. It indicates presumably the effect of a comparatively large local eddy, or vortex, which is combined with an overflow towards the sides. Thus the current styling trend to position the air inlet grills, for ventilation, at the base of the windscreen may be less practical when the angle between the windscreen and the bonnet has this critical value.

The flow visualisation of such an eddy with its axis parallel to the base of the windscreen and with side overflow is shown in the case of flat surfaces in Fig. 3.3.

The pressure distribution over the windscreen in the region of the re-attached flow measured along the plane of the vehicle's symmetry varies comparatively little with the change of angle γ. This is so in spite of the tendency of the airflow to spill over the sides. Fig. 3.5 shows the crosswise pressure distribution in the case of a flat windscreen. The pressure distribution over the bonnet in the crosswise direction and measured upstream of the line of separation is approximately the same in character as that over the windscreen. Plots of C_p/C_{pCL}, i.e. the ratio of the pressure coefficient C_p at a given point to the pressure coefficient at the centre line of the car were made for the flat surfaces of both the bonnet and the windscreen. This is of course only of comparative value as no modern vehicle now uses such styling. There are some bonnets however which are approximately flat. Most windscreens are curved and some are even curved three-dimensionally.

Consider now the effect of the windscreen curvature. A two-dimensional cylindrical curvature facilitates the flow sideways and thus tends to lower the pressure towards the sides. The windscreen curvature therefore helps to move the re-attachment line upstream, i.e. to lower it towards the bottom rim of the windscreen.

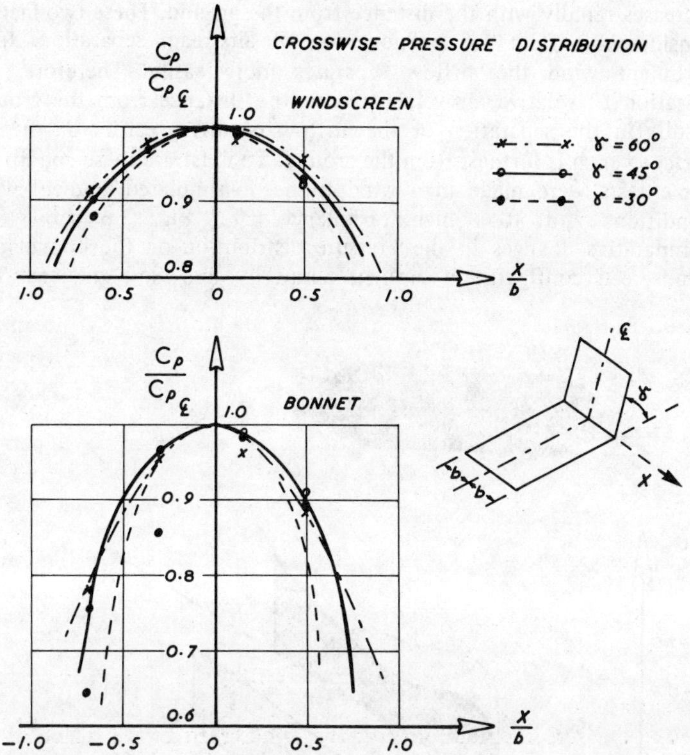

Fig. 3.5. The crosswise pressure distribution on the bonnet and windscreen configuration

It is perhaps reasonable to mention at this point that stylists who set the windscreen panel in a rubber surround, which constitutes a ridge around the windscreen, may cause the opposite effect, i.e. to delay the re-attachment of the airflow and move it downstream.

As the curvature of the windscreen facilitates the airflow spilling towards the sides, the shape of both the separation and the re-attachment lines will be affected. Both lines will tend to move towards each other, especially at the sides of the bonnet and the windscreen.

The interrelation of all the main factors involved in the analysis of the airflow pattern over the bonnet and the windscreen is deserving of more detailed experimental investigation.

The atmosphere is seldom at rest and any atmospheric wind introduces air turbulence and a vertical velocity gradient [18]. As a result of this vertical wind velocity gradient the relative air-velocity increases rapidly with the distance from the ground. These two factors considerably affect the phenomenon of airstream separation. In a turbulent wind the airflow separates more easily. Therefore, the variation of relative air velocity with the distance from the ground results in the separation of the airflow occurring earlier on the car surface which is furthest from the ground. To illustrate these conditions some tests were made in a wind tunnel in simulated gradient wind conditions and at a high turbulence rate. Fig. 3.6 shows the comparative changes in the pressure distribution on the bonnet and windscreen configuration brought about by the wind gradient. The

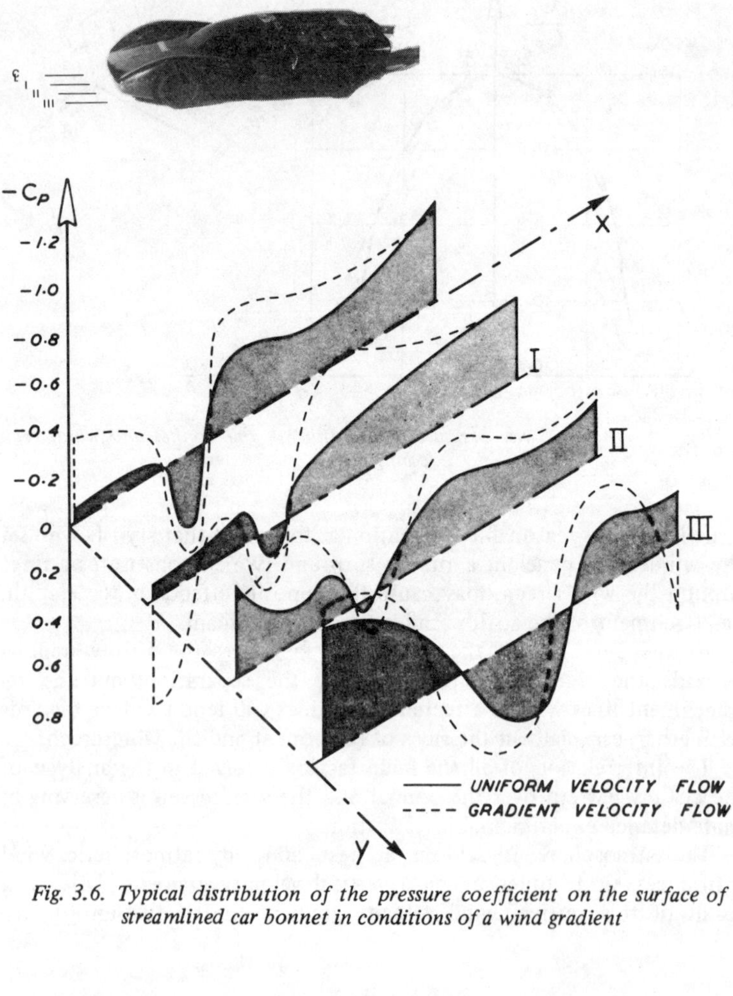

Fig. 3.6. Typical distribution of the pressure coefficient on the surface of a streamlined car bonnet in conditions of a wind gradient

continuous lines show the pressure distribution in the case of a uniform airflow while dotted lines indicate the pressure distribution in a gradient wind atmosphere. Road vehicles will often meet such gradient wind conditions and this effect therefore should be considered in car styling.

These remarks and descriptive or diagrammatic examples allow us to relate, from the point of view of the designer-stylist, the expected pressure distribution on the bonnet and on the windscreen. This study helps in selecting proper places for inlet grills, especially for ventilation, and the whole airflow pattern when visualized helps in understanding the behaviour of the aerodynamic force, particularly that part which affects the front steering axle.

The flow separation in front of the windscreen is a typical example of the effect of a sudden change in the flow direction. Flow separation in general is a complex phenomenon and the reader is referred to numerous textbooks on aerodynamics which analyse and elaborate on it in greater detail [19,20,21,22].

3.3 WAKE BEHIND A CAR

The main importance of airflow separation lies in the rear part of the car, where the separation line becomes the boundary of the turbulent wake behind the car. As we stated before, a moving car imparts its momentum to the surrounding air by means of viscosity. This volume of air forms a turbulent wake travelling behind the car. When driving behind another car we can often observe how dry leaves or snow flakes on the road surface are suddenly picked-up by the turbulence of the wake of the car in front and then travel with it along the road.

The phenomenon of the wake is best illustrated by a simple streamlined shape such as the axisymmetric body shown in Fig. 3.7. The flow visualization method shows a distinct circular line along which the flow separates from the surface. The wake is formed at the rear and surrounds the surface situated downstream of this separation line. The local pressure within the wake is practically uniform and below atmospheric. If we cut off the part of the body extending into the wake the actual resultant aerodynamic force acting on the car would not change. This operation of shortening the body immersed in the wake is colloquially called 'bobtailing' by stylists. This operation replaces natural separation by forced separation on the same line due to the sudden kink in the body shape, but the actual result is the same. For practical reasons the body must be terminated not far behind the rear wheel axle. Bobtailing is, therefore, an important aspect of styling and, as shown in Fig. 3.7, should theoretically be carried out along the plane

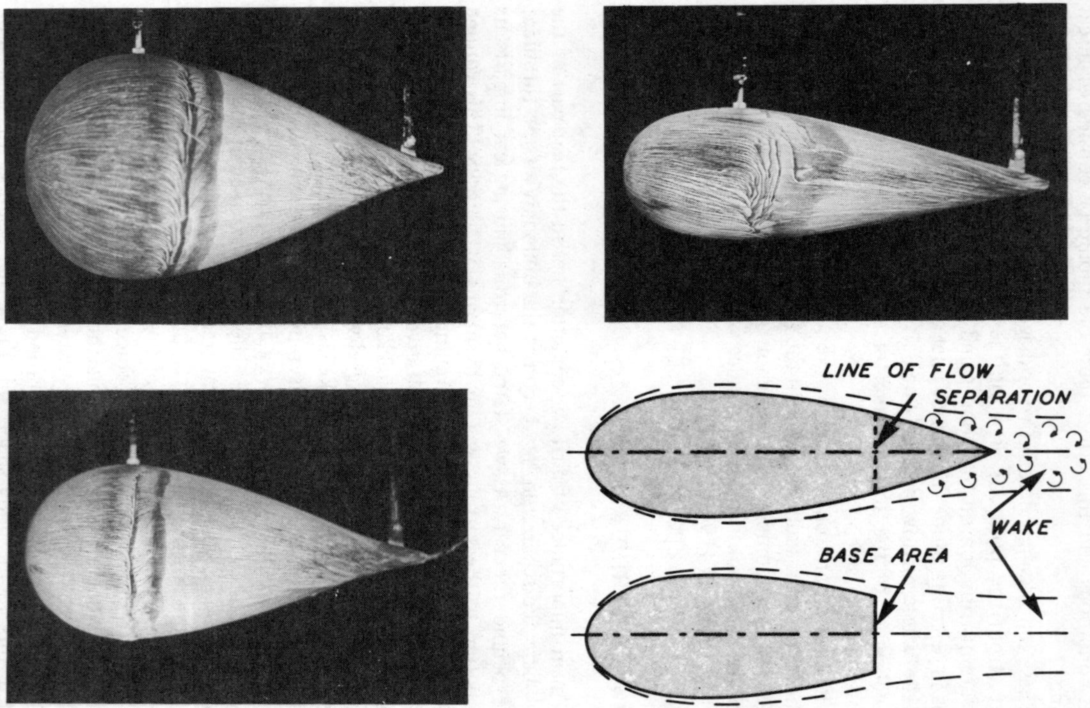

Fig. 3.7. Visualisation of flow separation on the surface of axisymmetric bodies and the principle of 'bobtailing'

normal to the main longitudinal axis of the car body. In the case of the axisymmetric body shown such a plane would cut on a circle at which the flow would in any event separate due to the flow conditions upstream.

The base drag

Consider now the general effect of bobtailing on the aerodynamic force. In the case of an axisymetric body, when its axis is parallel to the airstream, the aerodynamic force would be simply the aerodynamic drag. In the wake the negative pressure prevails so the rear cut surface is subjected to suction which produces a force directed against the motion of the body. We call this rear cut area the *base area,* see Fig. 3.7. It is therefore beneficial, with a view to reducing the aerodynamic drag, to shape the body so as to delay the natural separation as much as possible, to make the base area as small as possible and to keep the suction in the wake small. In other words we must try to maintain the highest pressure possible in the wake. There are some interesting methods which can be applied to this problem which will be mentioned later.

An actual car shape is much more complex than an axisymmetric body. It is made complex by the addition of a passenger compartment and the requirement of a window to the rear which influences the shape at the back. The actual line along which the flow separates differs greatly from car to car. Separation is often affected by design features such as the rear window frames, the positioning of water gutters, and other side vents. In the rear parts of the car, when the airflow is naturally slowing down because of the general negative slope of the external surfaces, any small protruberance can easily 'trigger' the flow to separate. Even a smooth rubber seal surrounding the rear window can upset the effect of a carefully styled rear shape.

The skin friction due to the flow over the part of the surface immersed in the wake is comparatively small as the relative air velocities are much reduced due to the tendency of the air mass to travel with the vehicle. A real and practical problem are possible dirt deposits obscuring the visibility to the rear. This problem requires careful study and an exact knowledge of the turbulence and eddy formation within the wake. Dirt is usually raised from the road surface by turbulence which due to the air mixing within the wake is deposited on to the rear surfaces of the car, including the rear windows. It is a matter of careful styling and design to avoid stirring the road surface by the turbulence of the wake in the first place and then to arrange the airflow in such a way as to induce it to wash the rear window and thus prevent dirt particles settling.

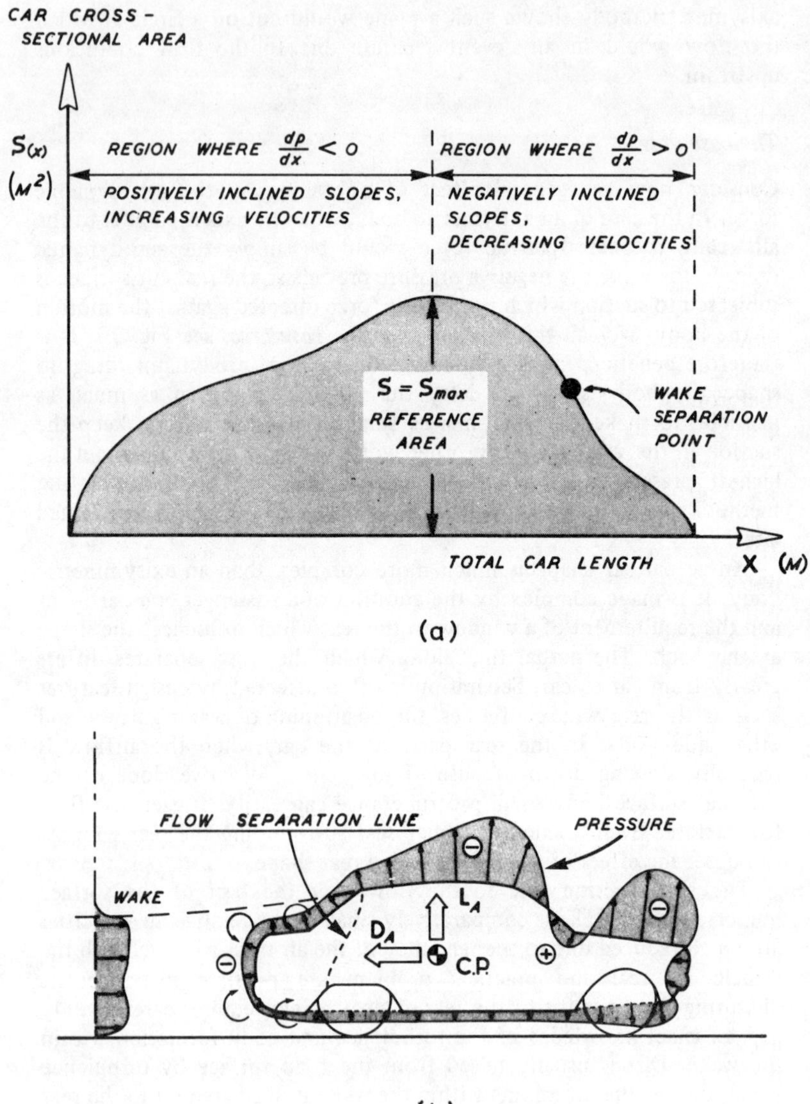

Fig. 3.8. (a) Lengthwise distribution of the cross-sectional areas of a car. (b) The wake and flow separation line related to the pressure distribution around a car

A sports car or an open racing car has a cockpit screened by a windshield. The airflow separates on the upper edges of this windshield which constitutes a separation trigger; the whole cockpit area is therefore immersed in a turbulent wake. In most sports and racing cars this wake is not re-attached to any other part of the car. In this way the base area includes a cross-sectional area which extends to the upper edge of the windshield. This concept of *base area* is convenient. When we integrate the prevailing pressure within the wake over this base area we obtain the *base drag*. In most cars the base drag constitutes an important part of the total aerodynamic drag and its reduction can greatly affect the performance. In the case of simple axisymmetric body shapes with a distinctive flat rear base this problem of a separate estimation of the base drag is quite straightforward. In the case of road vehicles which have much more complex shapes and are moving close to the ground the definition of the base drag may be vague. This is so because the rear wake behind the car often extends to the ground.

3.4 EFFECT OF WHEEL ROTATION ON AIRFLOW PATTERN

The airflow pattern around a car is complicated by the fact that it moves along the ground on wheels. By rolling on the ground the wheels rotate in an anti-stream direction. Points on the outer circumference of the tyre, when passing through the point furthest from the ground, are moving directly against the oncoming stream with double the car velocity, Fig. 3.9.

In most cars wheels rotate within cavities provided by the body design and are thus partially shielded from the effect of the outside airstream. In this case only the lower region of the wheel near the ground may be directly affected by the airstream. This of course reduces the influence of rotating wheels on general car aerodynamics.

To understand better the complex problems involved in wheel aerodynamics we should consider first the basic case of a simple wheel. Available experimental and theoretical material on this subject is still very fragmentary [23, 24, 25], and experimental research is continuing. The present investigation concerns itself with two interrelated aspects of this problem: first with the aerodynamic characteristics of the wheel itself and second with the influence of the wheel on the general flow pattern. Such an investigation can be carried out on scale models in wind tunnels.

Experimental investigations

An exact simulation of the rolling and flow conditions is technically difficult to achieve in a wind tunnel. In a wind tunnel air is blown on to

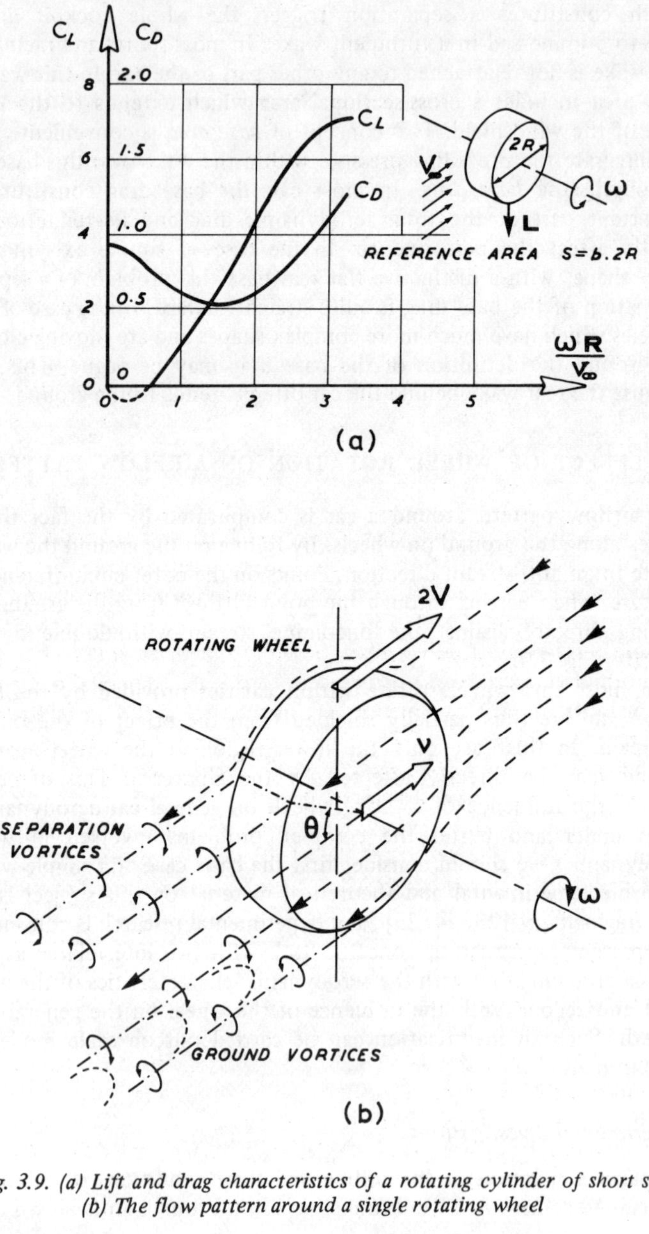

Fig. 3.9. (a) Lift and drag characteristics of a rotating cylinder of short span. (b) The flow pattern around a single rotating wheel

a stationary scale model, thus the real conditions are reversed. When the model is a wheel which is able to rotate along the ground the matter is more complicated. The scale model of the wheel should rotate touching the floor of the wind tunnel, which then has to move downstream with the same speed as the airstream. Such 'moving ground' facilities exist in some of the better equipped laboratories, but the fact that the wheel is in contact with the ground makes the force measurement almost impossible because instead of measuring the aerodynamic force we should be measuring the combined effect of the aerodynamic and the wheel-ground reaction forces. The moving ground is arranged in some wind tunnel laboratories by making the floor flexible and allowing it to move on two parallel rollers. Some laboratories, such as one in the University of Turin, have an arrangement whereby small electric motors drive large wheels which are in contact with wheels of the scale model of the car, Fig. 3.10. The electric motors *M* drive four large wheels *P* while the model is fixed over the dividing board, having its wheels *W* in contact with *P*. The required contact between the outer circumference of the wheel model and the simulated ground makes actual measurement of aerodynamic forces difficult in practice due to vibration and variable contact pressure. Some experimenters have rotated the model a small distance from the moving ground, but this introduces very intense local airflow augmentation in the gap between the wheel and the ground which influences the resultant aerodynamic force producing entirely false results.

Research on full-sized wheels exposed to the airstream (such as those on 'Formula I' racing cars) also present formidable practical difficulties, but mainly that of instrumentation. In such cases the wheel moves in the presence of the rest of the car and the interference between it and the body produces a very special flow pattern which is difficult to visualise and to investigate. The resulting flow pattern gives an accurate picture of the aerodynamic characteristics of the car, but as it is the product of many different influences, it is difficult to separate them out for the purpose of analysis.

Analysis of airflow pattern

With these problems in mind we return to the investigation of the wheel influence on the flow pattern over the car, having regard for the inadequacy of the available methods of measurement and taking account of all possible sources of error.

Consider the wheel as a cylinder of short span rotating in still air at a great distance from the ground, Fig. 3.9(a). Particles of air nearest to the surface of the cylinder adhere to its surface due to viscosity and consequently rotate with it. The viscosity of the air induces sur-

Fig. 3.10. (a) A single full-sized wheel and experimental test rig in a wind-tunnel (By courtesy of Prof. C. A. Morelli, University of Turin).

rounding air particles to follow the same path, which in effect creates the rotary motion of air around the wheel, and the airstream velocity distribution as depicted in Fig. 3.11(a). Such an airflow pattern is referred to in theoretical aerodynamics as a *vortex*.

The real airflow around the wheel is the result of superimposing two distinct airflows, the vortex and the uniform velocity airflow. Such a pattern is depicted in Fig. 3.11(a) and (b). The uniform wind is deflected by the rotation and the viscosity effect resulting in an asymmetric disturbance around the wheel. Therefore, the pressure distribution taken, for example, around the rim surface of the wheel is asymmetric which produces a resultant aerodynamic force. Such a force is the result of what is called the *Magnus effect* [20,21,22,26,27,28]†. It is directed at right angles to the oncoming airstream, i.e. in the case of wheels in the absence of the ground it would be directed downwards. This fact produces some confusion among stylists who argue that the wheel rotation creates a downward aerodynamic force. In fact since the wheels roll on the ground the airflow is restricted as air cannot flow

† Named after the German artillery expert, employed at the Court of the Russian Czars, who discovered the reason for the sideways motion of artillery shells due to their rotation. The same effect was utilised during World War II in the famous 'dam busting' bomb invented by Barnes Wallis.

Fig. 3.10. (b) Model racing car equipped with a rotating wheel system in a wind tunnel

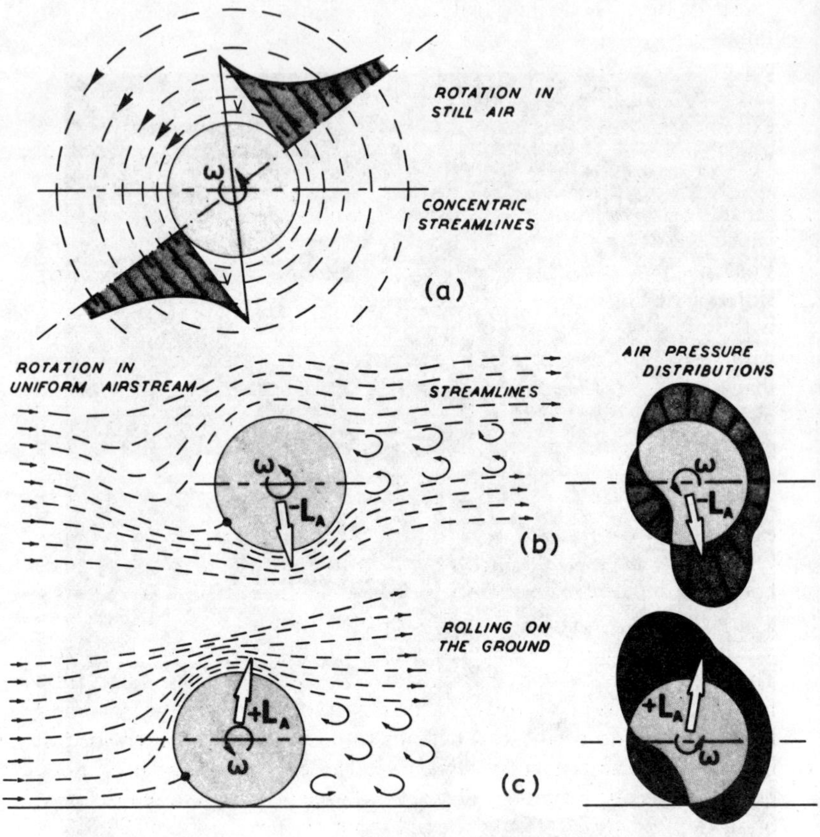

Fig. 3.11. Airflow pattern and pressure distribution around a rotating wheel in (a) still air (b) a uniform airstream, and (c) when rolling on the ground

between the wheel and the ground along the patch where the tyre makes contact with the ground. The resulting airflow pattern and corresponding air pressure distribution is depicted in Fig. 3.11(c). It shows that the resultant aerodynamic force acts upwards, i.e. it tends to lift the wheel from the ground. Thus we have in effect a considerable aerodynamic lift generated by the wheels. This is obviously true in the case of wheels which are wholly exposed to the airstream, like those in Formula I racing cars. When wheels rotate inside a more or less streamlined car body, like the G.T. sports cars or simple family saloons, the aerodynamic force generated seems to be fairly small and the airflow rather complex. The resultant aerodynamic force in this case seems to add to the drag rather than to the lift.

The model of a single rotating wheel was investigated in a wind tunnel [23]. In this experiment considerable problems arose in correctly simulating the conditions in which the wheel operates on a car. The model was first placed at a small ground clearance when extreme resistive pressures were measured in the region between the wheel and the ground due to the coinciding flow directions of the viscous surface layer moving with the outer rim of the wheel and the general airflow. This suction produced a downward aerodynamic force which in the case of wheel-ground contact would be reversed. Consequently the measured forces did not represent the true situation. Other measurements made on scale models of a racing car with and without rotating wheels have shown that measurements with stationary wheels may give an aerodynamic lift of about 30% less than those with their wheels rotating [23]. As is usual in wind tunnel measurements, the wheels were stationary in order to reduce the expense of model making and these results should be taken with reservation. This would be especially true for cars with wheels exposed to the airstream. The Formula I racing car has a body with four fully exposed very large wheels. Thus the rotating wheels create a complex velocity stream pattern, interfering with the airflow over the main body. An analysis of this interference and an assessment of its effects is a very complex matter and requires further study.

The single rotating wheel induces a vortex flow, as already stated. This air mass induced by the viscosity to rotate with the wheel discs forms a weak form of vortex. When this vortex meets the oncoming flow on the side of the car, it tends to be bent down. These vortices interact with the general system of vortices created by the main body of the car and are eventually absorbed. At the same time when the mass of air induced to move with the outer rim reaches the top of the wheel it meets the oncoming airflow which works directly against it. As the air at the upper surface of the wheel travels at about twice the car speed (in still air) it follows that at the upper part of the rotating wheel particles of air still travel in the direction of the car motion. As a result the airstream velocity in this region decreases rapidly as the mass of air piles up against the oncoming stream. Therefore we should expect to find somewhere upstream of the top of the wheel a point at which the airstream separates from the tyre (Fig. 3.12(a)). This separation point is clearly dependent on the speed of the car and moves forward as the car speed increases. This phenomenon can be studied in a wind tunnel by flow visualization methods, such as filming the behaviour of woollen tufts placed in the proximity of the wheel surface, or by measurement of the local air pressure. This problem of flow separation is illustrated in Fig. 3.12(b).

Racing car enthusiasts, when watching cars on the track on a rainy

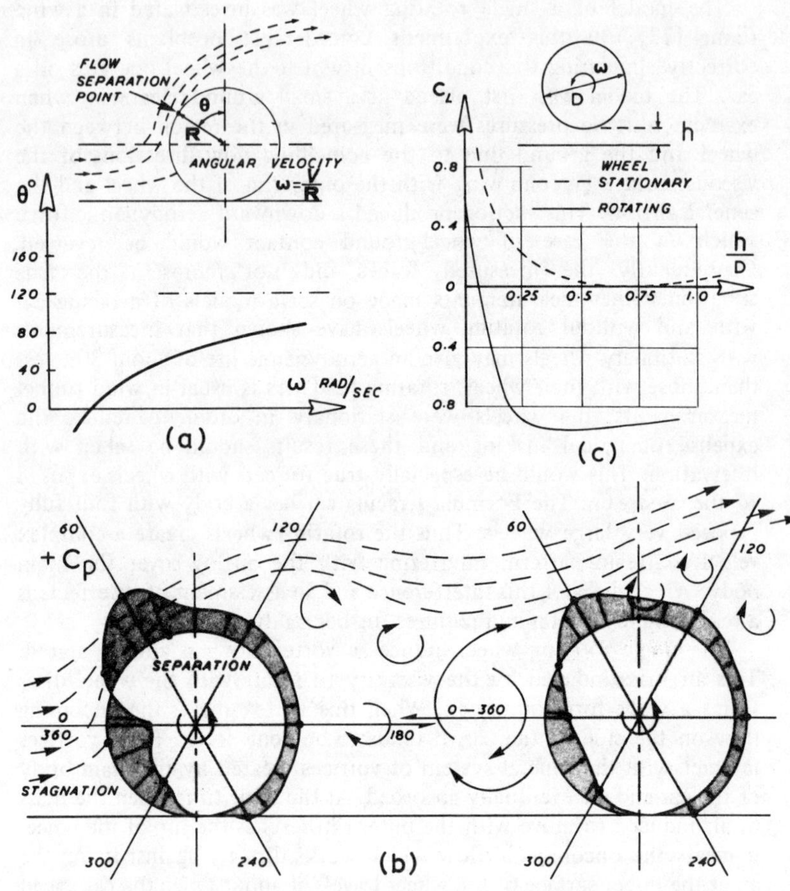

Fig. 3.12. (a) Flow separation on the surface of a rotating wheel; (b) airflow pattern around wheels rolling in tandem; (c) 'magnus effect' corrected for the ground presence for a wheel rotating just above the ground

day, can see a turbulent wake formed over the upper part of the body. This wake carries pulverised water droplets which seem to originate somewhere in front of each wheel. The aerodynamic drag force is considerably influenced by this phenomenon which occurs with exposed wheels. It slightly reduces the aerodynamic lift component but adds to the horizontal drag component. The exposed wheels of such racing cars contribute greatly to their air resistance, so that wheel drag constitutes up to about 45% of the total drag of the vehicle.

Further research work has been conducted over a number of years in the wind tunnel of Pininfarina at Turin [88]. One of the problems which has been investigated is the effect on the measured loads on a wheel due to the small clearance gap which is often left under a rotating wheel when it is set up in a wind tunnel on a fixed floor. The results showed that on the ground directly beneath the wheel the pressure decreased steadily as the gap diminished. Minimum pressure coefficients of about −1.80 under a rotating wheel and about −1.67 beneath a stationary wheel were observed at gaps of between 1% and 1.5% of the wheel diameter. However, when the wheel was in contact with the ground the pressure coefficients immediately ahead of the contact patch were in the region of +1.0. This clearly shows the very rapid change in local pressures and hence in lift and drag forces which must occur with change of gap height at the small clearances which have been used in wind tunnel investigations of the aerodynamic characteristics of wheels.

Cogotti and his colleagues found that the most effective method of mounting wheels for such tests was to use polystyrene foam and foam rubber pad inserts between the tyre and the floor of the wind tunnel or the balance, the pads having an area equal to that of the tyre contact patch under typical road-going conditions. However, some doubts still remain when this method is applied to tests involving typical treaded tyres as are used on most saloon cars because there may be air passage between the treads in the tyre contact patch which will have an effect similar to that of a small clearance gap.

The Pininfarina tests also showed that there were dramatic changes in the drag and lift coefficients of an isolated tyred wheel in contact with the ground as the wind speed increased, or more accurately as the size and speed parameter known as the Reynolds number increased.* The drag coefficient reduced from about 0.60 to 0.52 and the lift coefficient reduced from 0.70 to about 0.35 for a stationary wheel having a tyre with width to diameter ratio of 39%, and there were detectable differences dependent on whether an open, standard rim alone was used or whether it was faired with a flat cover. These changes in force coefficient, which occurred as Reynolds number increased from 10^5 to 10^6, are similar to those which are observed in flow at similar Reynolds numbers past circular section cylinders, spheres and other shapes with rounded corners. In these conditions the position at which the flow separates from the surface of the body is dependent on a number of effects which include the roughness of the body surface and the amount of turbulence in the approaching airflow.

* In air, the Reynolds number $R_e = 69{,}000\ V_A l$ where V_A is the relative wind speed in m/s and l is a characteristic length dimension of the body in question measured in metres.

For a tyre of 0.5 m diameter, a Reynolds number of 10^6 (i.e. 1 million) is achieved at a relative airspeed of about 30 m/s or 66 m.p.h., and since different tyre tread patterns have different degrees of roughness and the flow into a wheel cavity is often highly turbulent, it is clearly important that wheel and tyre tests should cover a suitable range of relative wind speed so as to ensure that data is available at the correct Reynolds number for use by the car designers. Some of the earlier wind tunnel tests on models of wheels could not meet this requirement, but the results which they generated are qualitatively correct.

The effects of wheel rotation on an isolated wheel at full scale and with various clearance gaps have also been investigated. These tests showed that with no ground clearance gap the drag coefficient of both the stationary and rotating wheel was nearly 0.60 whilst the lift coefficient for the stationary wheel was 0.27 and for the rotating wheel it was 0.18. The force coefficients on both the stationary and rotating wheel changed rapidly as the ground clearance opened up to about 2% of the wheel diameter, the drag coefficient reduced by about 0.05 for both situations whilst the lift coefficient reduced by about 0.3 for the stationary and rotating wheel, going negative for the latter. When the clearance was increased from 2% to 10% or more of the wheel diameter both the drag and lift coefficients were almost constant, the stationary wheel having values of drag and lift coefficient of about 0.50 and 0.03 respectively, whilst the corresponding values for the rotating wheel were about 0.55 and −0.15. When nearly-flat fairings were fitted to both sides of the wheel in contact with the ground the stationary wheel experienced a reduction of 0.05 in drag coefficient and an increase of 0.02 in lift coefficient, whilst for the rotating wheel the force coefficients reduced by 0.09 and 0.002 respectively.

Cross-flow through plane of wheel rotation

Aerodynamic drag is also caused by another phenomenon related to wheel rotation. The velocity distribution in the axial direction shows that the flow on both sides of the wheel is often locally slower than the flow over the car generally. Slower flow corresponds to higher local pressures (see Equation 2.1). It is a physical fact that air tends to flow from regions of higher local pressure to regions of lower local pressure. Thus in the general airflow pattern we have to consider the cross-flow effect. Cross-flow combined with the overall flow due to the car motion results in a vortex. When we consider a single wheel we have such a pattern in two regions, at the flow separation line (Fig. 3.9(b)) and near the ground. When a car travels along a dusty road it is possible to see the effect of the vortices created by the wheel near the ground. Careful study of the marks made by dust on the road also show the sign of the side vortices parallel to the pattern of the tyre imprint.

The creation of all these vortices requires energy which is dissipated from the wheel momentum. This constitutes part of the wheel drag. It is difficult to measure this contribution separately from the total wheel drag.

In a Formula I racing car the wheels are for all practical purposes arranged in tandem. The wheels rolling at the rear move totally or partially in the wake of those at the front. Thus the rear wheels move in different conditions and accordingly their aerodynamic behaviour is different. The general pattern of such wheel interference is depicted in Fig. 3.12(b). It is generally accepted from available experimental data that the front wheels in these cars contribute approximately 65% of the total wheel drag.

It is also known [24] that the drag and lift forces acting on a wheel depend on the yaw angle and that they increase from minimum values when the plane of the wheel is in line with the approaching wind flow direction, to local maximal values when the yaw angle reaches about $\pm 15^\circ$. However, the side force, yawing moment and rolling moment about ground level all increase steadily in direct proportion to the yaw angle for angles up to about 30°. The sensitivity of these force coefficients to the local flow yaw angle can be detrimental to good aerodynamic design. The flow pattern beneath a car will be discussed in detail in Section 3.5 but it is here sufficient to state that the local flow on the underside of a car may be approaching the wheels with significant deviation from the original flow approach direction. Not only will this give rise to an increase in the aerodynamic drag of each wheel so affected, but it could well generate a significant interference drag increase by modifying the pressure distribution within the wheel cavity and by changing the pattern of the air flow out of the wheel cavity. Front wheels are likely to be more severely affected by these cross-flow effects than are rear wheels, and so great care should be paid to the location of wheels and wheel layout arrangements if a low drag configuration is required.

One such project researched at Pininfarina [88], using a model of a low drag body shape with fairings around the wheel cavities and flat covers over the wheel rims to bring them flush with the tyres, showed a drag coefficient of 0.157 when the wheels were mounted conventionally. When the wheel layout was changed to a diamond pattern with a single central front wheel, a pair of conventionally-situated rear wheels in faired cavities and a centrally mounted 'tail' wheel, the drag coefficient was reduced to 0.134. Flow visualisation tests had shown large airflow direction changes over the floor of the wind tunnel near the front of the model with the higher drag value.

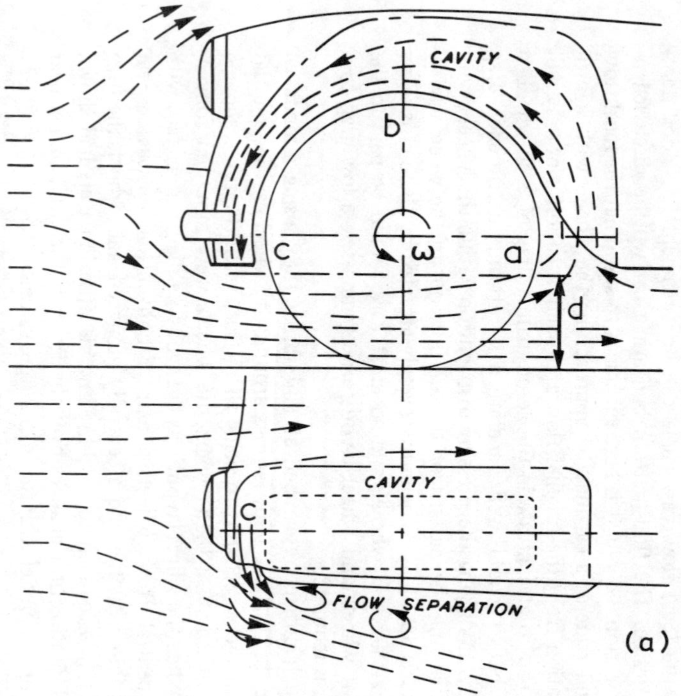

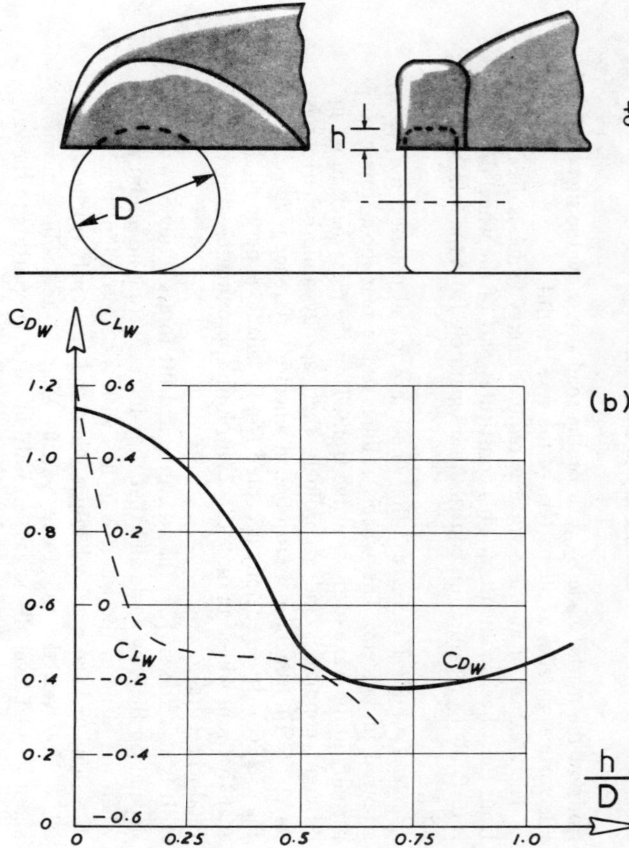

Fig. 3.13. (a) Airflow within the front wheel cavity; (b) mudguard effect on wheel lift and drag

Airflow inside a wheel cavity

Racing cars are in many respects exceptional and we should be more concerned with saloon car shapes. We know little about the airflow behaviour around a wheel rotating inside a cavity formed by the car chassis and the mudguard. The only region through which air mixes with the airflow within such a cavity is region *d* (Fig. 3.13(a)) near the tyre-road contact which is limited by the car underside and ground clearance. It may be assumed that the wheel rotation in such a confined space would, by the mechanism of viscosity, tend to suck in air through area *a* at the rear of the cavity and carry it around through region *b* towards region *c* in front of the wheel. The external airstream which moves through the underbody and the ground is somewhat different in the region of the front wheels than in the region of the rear wheels. In the region of the front wheels the flow underside is still comparatively little affected by the formation of the viscous boundary layer on the underside of the car, or by different protuberances on the usually uneven car underside surface. The airflow there tends at first to speed-up locally between the lower lip of the car front and the ground. We may therefore assume that the local air pressure would be lower than in region *c* within the cavity. This induces the air to squirt out from the cavity, and to mix with the outside airstream. That this is so may be observed when passing other cars either on a rainy day or when there are dry leaves on the ground. The efflux of air from the wheel cavity creates conditions of turbulence which affects the rear part of the front wheel arch. It is often visible on a muddy day when the part behind the front wheel arch is covered with a dirt deposit. This is due mostly to the flow separation induced by the air blowing out from the cavity at the front of the wheel.

All this affects the momentum of the car, as the kinetic energy of the car is dissipated by air friction and turbulence. Experimental research has shown [29] that gradually decreasing the clearance between the underside and the ground and increasing the proportion of the wheel rotating within the cavity tends to decrease the total aerodynamic drag coefficient of the wheel, C_{D_W} (see Fig. 3.13(b)). The values of this coefficient were calculated for the wheel frontal area as the reference area, as given in equation 2.2. This decrease is understandable as the total exposure of the rotating wheel to the airstream involves considerable loss of momentum to the air which the partial shielding from the external airstream tends to reduce. It was found also that a further increase in shielding of that part of the rotating wheel in the cavity causes the viscous losses to predominate. This is shown by the tendency of C_{D_w} to begin increasing after reaching its minimum. The minimum wheel drag coefficient is reached at a shielded height to wheel diameter ratio of about 0.75.

The curve showing the variation of the aerodynamic lift component, Fig. 3.13(b), shows a positive lift when the hood over the wheel is fixed, i.e. when $h/D = 0$. The reason for this was explained earlier in this section. When the mudguard is gradually lowered the values of the lift coefficient drop and become slightly negative. As the results for large values of h/D are taken from wind tunnel tests [29] which were carried out on a model of a wheel with false ground contact simulation they may be grossly inaccurate. (To allow rotation the outer rim of the wheel, instead of touching the ground, dipped into a cavity cut out in the rigid board representing the ground.) It is probable that through this slot the fast moving airstream created a powerful suction towards the ground which reversed the sign of the aerodynamic lift. Further investigation shows [29] that careful exclusion of this gap and a more correct simulation of the real airstream conditions produces a positive sign for the aerodynamic lift force.

The steering wheels require cavities wider and deeper than those of the rear wheels. These cavities therefore are more open to the outside flow and as the pressure difference is more acute at the front, the flow outwards interferes more with the total airflow pattern. The shielding of the frontal side openings tends to prevent this outflow and consequently reduces this effect. Statistical values of total drag based on scale model experiments in wind tunnels (in which, admittedly, the rotation of wheels is usually not simulated) indicate that for an average saloon car the reduction of drag coefficient C_D achieved by shielding is about 3.5% [30]. The effect of wheel rotation would probably cause a further reduction of about 5%.

Rear wheels are often substantially or even completely shielded. The outflow of air from the rear wheel cavities is usually less than occurs at the front primarily because the pressure gradient between the inside and outside flows are smaller. Consequently the resulting rate of outflow does not affect the external flow pattern so much, and being at the rear covers a comparatively small part of the vehicle.

In summary the outflow from the rear wheel side openings add little to the overall drag of the car. Statistics available [30] indicate that total shielding of the rear wheel side openings could reduce the overall drag coefficient by approximately 1%.

A detailed investigation [88] has been made into the effects of width and height of a wheel cavity on the drag and lift of a drop-shaped body of low drag which encased two full-sized wheels and a driving motor. These tests in the Pininfarina wind tunnel showed that the lift of the body was more sensitive than the drag to change in the ratio of the cavity width to wheel width and to change in the ratio of cavity radius to wheel radius. The drag coefficient increased by about 10% from a value slightly above 0.17 in both situations whilst the lift coefficient increased by almost 50% from an initial value of about 0.13. Perhaps the more significant result was achieved when the changes in the two geometric parameters were combined into the single ratio of the volume of the cavity to the volume of the wheel. The increase of drag coefficient was still small for the corresponding increase in volume ratio whilst the relatively large changes in lift coefficient correlated well with the increase of volume ratio. Changes in drag and lift coefficient resulting from having the wheels stationary or rotating were much smaller than those which resulted from changes in the geometric parameters of the cavity.

The extension of these results to full scale motor cars should be treated with a little caution, but the benefit of small wheel cavities should be retained even if the details of the flow are slightly different and dependent on the shape of the particular car under consideration.

The effect of the wheel width

There is a marked tendency towards the use of tyres of increasing width—a 'fashion' identifying with the speed concept of racing cars. Wide tyres are used, understandably, on fast sports cars, and racing cars use exceptionally wide tyres to improve the tyre-road interaction. For a given wheel diameter a wide tyre results in a change in aerodynamic characteristics. It is as if we consider a rotating cylinder of gradually increasing span. It ceases to be a wheel when the diameter is less than

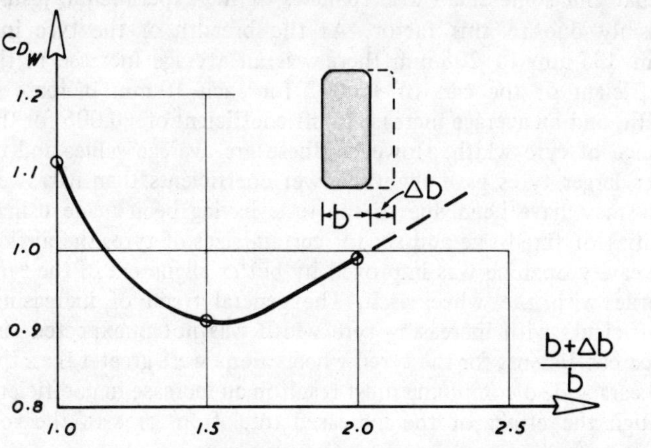

Fig. 3.13. (c) Effect of wheel width on aerodynamic drag

its span, or width, and becomes a cylinder when the diameter to span ratio is of order of unity.

Preliminary investigations of a single wheel exposed to the airstream [24] have shown that increasing the width produces different aerodynamic drag characteristics. The variation of drag coefficient with the width of the wheel is shown in Fig. 3.13(c). Considering a normal tyre width b and increasing it gradually by an additional width Δb produces at first a decrease and then an increase of the drag coefficient. The curve of C_{D_W} in terms of the ratio $b + \Delta b/b$ is a minimum when the width increases by about 60%, but this depends on the width to diameter ratio of the initial tyre configuration. Any sports car enthusiast who contemplates an increase of speed by installing extra wide tyres should bear this in mind. Using extra wide tyres not only increases the total tyre rolling drag but also is most likely to increase the aerodynamic drag due to the tyres themselves, and can increase their aerodynamic interference with the flow pattern around the wheel cavities and the car. The improved traction of the driving wheels can be utilised to increase acceleration at middle-range speeds, but will not result in increased maximum speed due to the limited engine power available, and the net result here may be negative.

Systematic tests have been made where a number of sets of tyres of increasing width and almost the same diameter were fitted to two different cars and the drag and lift coefficients were measured in the Pininfarina wind tunnel with the wheels stationary [88]. Change of frontal area used in defining the force coefficients (see Section 2.2) was taken into account but the wheel rims used were not perfectly similar and some small discrepancies in the experimental results were possibly due to this factor. As the breadth of the tyre increased from 135 mm to 205 mm there was an average increase in the drag coefficient of the cars of +0.002 for each 10 mm increase in tyre width, and an average increase in lift coefficient of +0.006 for the same change of tyre width. However, these are average values and in some cases larger tyres gave slightly lower coefficients than narrower ones. This may have been due to the tests having been made using wheel cavities of fixed size and so, for certain sizes of tyre, the airflow over the cavity opening was improved by better alignment of the tyre outer profile with the wheel arch. The general trend of increasing force coefficients with increasing tyre width was not unexpected since the force coefficients for the tyred wheels alone were greater than those for the cars and so combining must result in an increase in coefficient, even though the effect of the increased total front area of the vehicle is offset.

One of the cars used was also retested with a front air dam fitted, which reduced the drag coefficient by about 0.125 and reduced the lift coefficient by about 0.025, but did not change the effects of tyre width on the drag and the lift coefficients.

The use of wider, low profile tyres has become widespread on current saloon cars. These tyres have lower rolling resistance and improved road holding characteristics, but tend to have higher aerodynamic drag than the narrower tyres they have replaced. It is not yet clear whether the increased drag is due to their greater breadth to diameter ratio, or to their increased frontal area, or to aerodynamic mismatching of the rotating wheel to the car body.

The fitting of flat or nearly flat wheel fairings over the rim of car wheels to give a smooth shape to the outer surface of the wheel has been known to reduce the drag of a car. Statistical analysis of test data by Adam Opel A.G. [88] has shown that the average effect on fitting to production cars with standard wheel rims is to produce a reduction of vehicle drag coefficient by about 0.009 for stationary wheels. This is generally, but not always, accompanied by an increase in the lift coefficient of the vehicle by a similar amount. Often small vents have to be introduced into such covers to allow through-flow of air to cool the brake discs and pads. Such wheel fairings are increasingly common fitings on many medium size executive-type cars of British or Western European manufacture.

3.5 AIRFLOW BETWEEN UNDERSIDE AND GROUND

The motion of a car on the ground introduces problems which differ greatly from those of an aircraft, primarily because of interference with the airflow between the car underside and the ground. As the car moves very close to the ground this interference is one of the most important features of the airflow pattern around it.

The airflow around any shape when it is moving in an unobstructed free atmosphere is different from the airflow around the same shape at the same attitude when it moves in the proximity of the ground. Consider as an example the simplest type of aerofoil. A symmetric aerofoil moving in a free atmosphere at zero incidence has a symmetric pressure distribution on the upper and lower surfaces and consequently it will have a zero aerodynamic lift component. Such an aerofoil is then subjected only to the aerodynamic drag force resisting its motion, Fig. 3.14(a). The fact that there is no aerodynamic lift is due to the symmetric airflow pattern. In the proximity of the ground (Fig. 3.14(b)) the airflow pattern changes drastically and becomes asymmetric with respect to the chord. This produces in turn an asymmetric distribution which when integrated around the contour of the aerofoil gives an aerodynamic force. This force is inclined to the

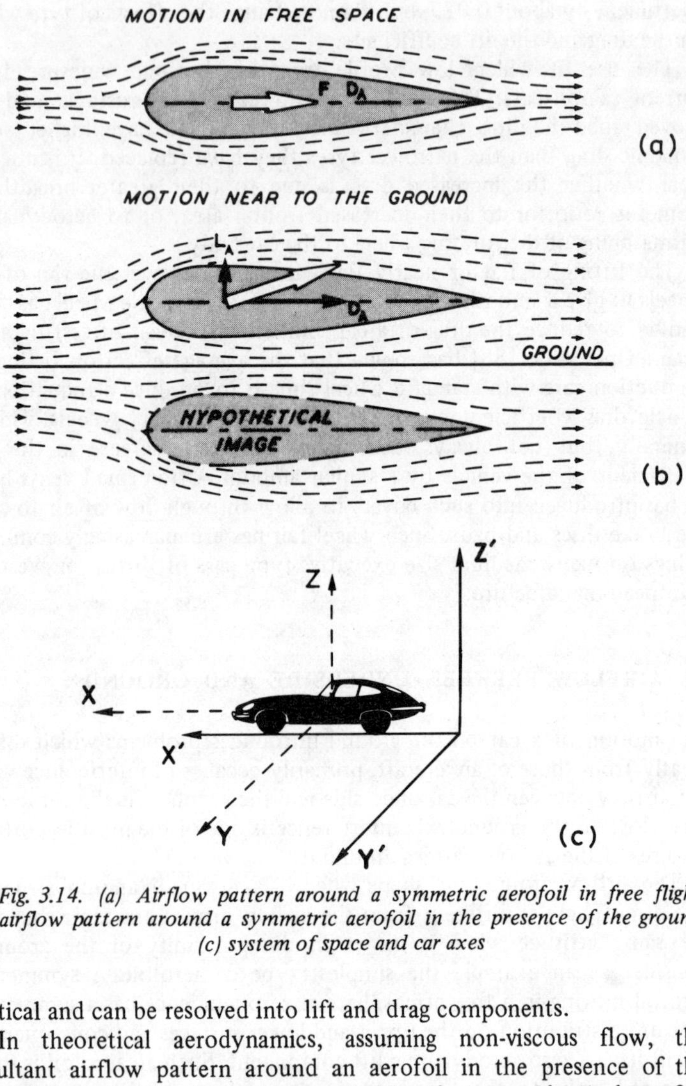

Fig. 3.14. (a) Airflow pattern around a symmetric aerofoil in free flight; (b) airflow pattern around a symmetric aerofoil in the presence of the ground; (c) system of space and car axes

vertical and can be resolved into lift and drag components.

In theoretical aerodynamics, assuming non-viscous flow, the resultant airflow pattern around an aerofoil in the presence of the ground is similar to the pattern created by placing an identical aerofoil symmetrically to the ground surface but on the opposite side. This is known as the 'principle of images'. Some wind tunnel laboratories use this principle to simulate the presence of the ground but the method is only an approximate one because it assumes the air to be non-viscous. In the case of vehicles which move very close to the ground the total distance between the model and its 'image' in the wind tunnel would be comparatively small and the effect of air viscosity could affect the results. However, bearing this in mind, the local airspeed between the aerofoil and the ground is increased so that there is an increase in negative pressure and the aerofoil is therefore attracted to the ground. This is called negative lift, an effect well known in aerodynamics [28], and has been theoretically and experimentally investigated by many researchers [31,32,33].

Theoretical calculations, in the extreme case when the aerofoil almost touches the ground, produces (neglecting viscosity) a maximum negative lift coefficient of the order of −0.37. In practice this theoretical case is complicated by the fact that the real airflow is viscous and consequently the assumption of non-viscous flow is not valid when the vehicle is in close proximity to the ground. This viscous effect can result in the aerodynamic force changing sign and becoming positive, i.e. directed upwards away from the ground.

Let us consider a case in which the vehicle moves in a still atmosphere, and let us assume that the underside of the car consists of a single smooth panel. In this case the relative velocity of the air and the car is the same as that of the car and the ground, Fig. 3.14(c). Fixing one set of axes x, y, z, to the moving car and a set of references axes x', y', z' to the ground we may say that the x, y, z system moves with a given relative speed V with respect to the x', y', z' system.

We stated previously that the disturbance to the airstream created by the motion of the car produces local accelerations and decelerations. These in turn cause the local air particles to have, temporarily, a relative velocity with respect to the ground. The relative airflow causes the development of the viscous boundary layer on the outer skin. Such a boundary layer, therefore, forms on the underside panel of the car, just outside the stagnation region where air moves relatively slowly [15,16].

The case of viscous flow between two infinite surfaces, one moving relatively to the other (Couette flow), has been investigated theoretically in aerodynamics but this theory is of little help here due to the relatively limited width and length of the car's underside which makes the viscous flow there highly 'three-dimensional'.

From the point of view of an observer moving with the vehicle, the boundary layer is the layer of air which tends to move with the car. The air disturbed by the passage of the car tends to increase in speed. Thus, although initially at rest with respect to the ground (no wind), air would begin at first to move over the ground in a direction opposite to the car motion. Then, when the air, due to viscosity, adheres to the car it tends to travel with it (in the boundary layer and in the wake) reversing the local air motion relative to the ground which is then in the direction of the car motion. Thus viewed from a point on the ground, the air immediately above tries first to move backwards and then forwards when the car passes over it. This local gust often produces a 'whirl' which travels along the ground together with the car–a sort of 'separation bubble'. The air seems to be rotating within it, as shown by arrows in Fig. 3.15.

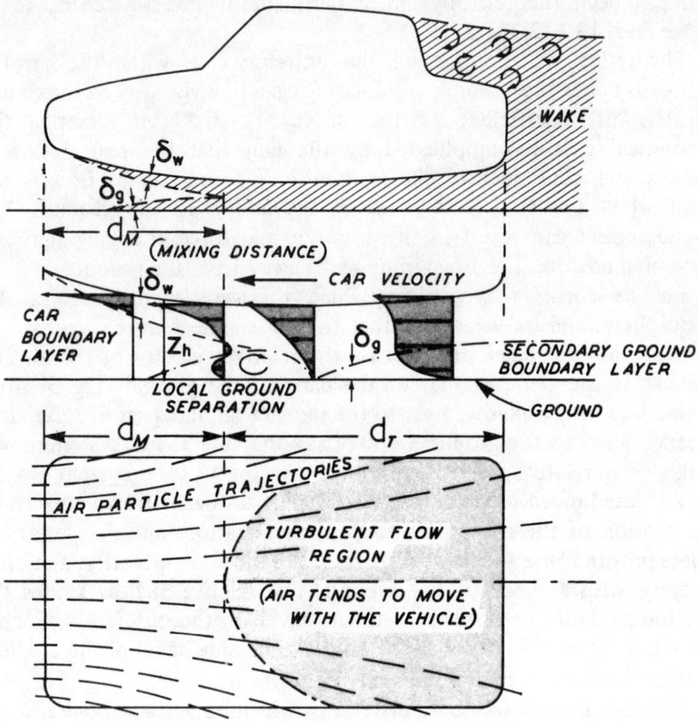

Fig. 3.15. Airflow pattern underneath a car

The car boundary layer thickens in the downstream direction and depending on z_h, the actual total ground clearance (of the order of 15 cm), it may reach the ground. In this case it will merge with the rear wake which will then reach to the ground. When this boundary layer touches the ground the distance from the front of the car to the point at which it impinges on the ground is called the *mixing distance,* d_m. Further downstream from this point the whole mass of air between the underside panel and the ground is induced to travel together and slides along the surface of the ground. This in turn creates a condition of relative velocities which leads to the formation of a *secondary* boundary layer.

We see that the whole flow pattern on the underside of the car is comparatively complex. In general, the flow there is highly turbulent and this secondary layer adds to the turbulence. The flow pattern produces a pressure distribution which, when acting on the large planform area of the underside panel, greatly contributes to the aerodynamic lift. The control of this pressure distribution is then an important matter.

Effect of ground clearance and smoothing the underside

In many cars where the engine is fixed behind the front axle, the underneath of the engine compartment is open to the ground to improve the cooling of the engine's crankcase. Thus a cavity is formed full of structural or suspension members which may produce flow separation just behind the front bumper. Air inside this cavity is usually affected by the front grille and the high dynamic pressure region in front of the vehicle. This helps to create an additional aerodynamic lift force. The damaging effect of such a lift is discussed later.

Most present road vehicles have very rough undersides (only a few cars have an underside panel) due of course to the presence of exhaust pipes and silencers, overhung petrol tanks, wheel axles, differentials, suspension shock absorbers, brake cables and wheel cavities, etc. The average roughness is ± 15 cm (or 6 in) considered from a 'mean surface' level. Any boundary layer adhering to such an underside can simply be considered as a volume of air trapped underneath the car while around the perimeter local velocity gradients cause the formation of eddies and turbulence.

The airflow between the underside and the ground is therefore affected by

(1) The distance (clearance) between the ground and the underside.

(2) The width, length and height ratios of the vehicle and the styling of the body shape.

(3) The roughness of the underside.

(4) The lengthwise and crosswise curvature of the underside panel.

The ground clearance z_h depends on the design and geometry of the car. Any acceleration or deceleration of the car causes a change of ground clearance due to its elastic suspension. It is obvious that any change in the clearance can considerably effect the airflow underneath and consequently the flow pattern around the car. If we were to put a soft rubber curtain around the underside panel in such a way as to exclude any airflow (but not restricting the car motion mechanically) we should then travel with a quantity of air trapped inside at atmospheric pressure. At the other extreme we can increase the ground clearance to its maximum practical value which will present the minimum obstacle to the airflow on the underside and lower the pressure. It is reasonably safe to conclude that a free flow underside would considerably reduce the mass of air which would normally be flowing over the upper surfaces of the car. Therefore the general flow pattern around the car depends on the ground clearance. The effect of z_h on the values of the drag coefficient in the case of smooth and rough undersides is shown in Fig. 3.16(a). Fig. 3.16(b) shows an example of a car that has $C_D = 0.35$ at its normal ground clearance.

The effects of ground clearance would be different for very rough or smooth undersides. Some researchers suggest that a very rough underside and a large clearance, as in trucks and lorries, actually causes an increase of drag by creating conditions of greater freedom for the formation of eddies in the underside flow [34]. On the other hand, a general increase in ground clearance leads to a more unobstructed airflow which by bleeding air from other flow regions produces, in effect, a decrease in the aerodynamic force. Thus the general effect of smoothing the underside will be different for cars of good and bad aerodynamic shape. Fig. 3.16(c) shows diagrammatically the effect of the ground clearance increase on the car lift coefficient. These diagrams were produced for cars with an average underside roughness and show that vehicles with bad aerodynamic styling, characterised by large total drag coefficients, display a slight increase in drag while those of good aerodynamic shape display a rapid decrease in drag with increasing ground clearance. The smoothing of the underside also reduces the amount of turbulence and frictional losses. This contributes to the reduction of the aerodynamic drag.

The air trapped in the underside region tends to slow down the flow and consequently increases the ambient pressure there, while the airflow along the sides, which is comparatively unobstructed, moves faster and the ambient pressure there is consequently much lower. As the air always tends to flow from regions of higher to regions of lower pressure the air flow underneath the car tends to turn sideways. This affects the general flow pattern which becomes much more 'three-dimensional'. The sideways airflow from the underside of the car is often quite visible to the observant driver following a car along a road covered with dry leaves.

The atmospheric wind which produces a velocity gradient across the road may affect the general airflow pattern over the car but it can be argued that it would not much affect the flow underneath as the wind velocity would be very small close to the ground. A more detailed analysis of the effect of ground interference on the aerodynamic lift is presented in the next chapter.

Problems connected with the underside cannot be considered separately from the rest of the vehicle shape. The airflow around complex shapes such as vehicles produces different degrees of inter-

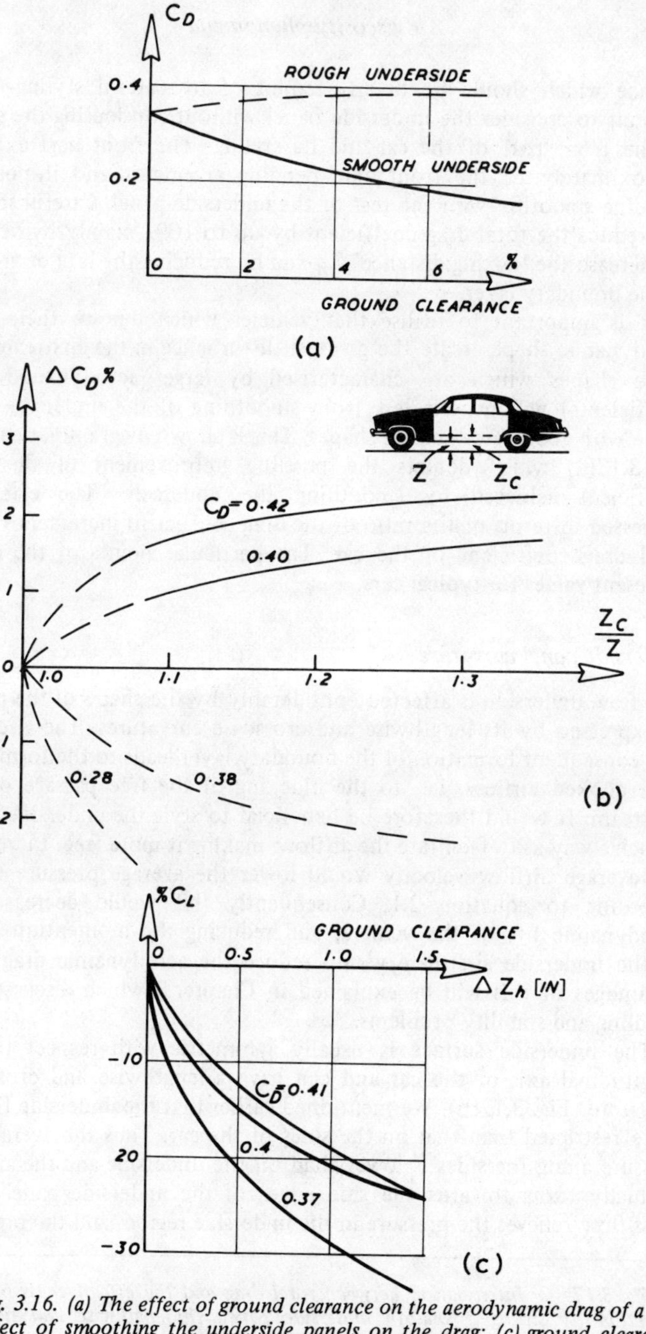

Fig. 3.16. (a) The effect of ground clearance on the aerodynamic drag of a car; (b) effect of smoothing the underside panels on the drag; (c) ground clearance effect on the aerodynamic lift

ference which should be taken account of by careful styling. It is difficult to consider the underside panel without considering the shape of the front part of the car and its styling. The front part extends approximately to the front axle bending around it and if possible blending smoothly with the rest of the underside panel. Careful styling can reduce the total drag coefficient by up to 10%, mainly by helping to increase the 'mixing distance' d_m and by reducing the rate of growth of the boundary layer.

It is important to realise that vehicles which due to their poor aerodynamic shape create the greatest disturbance in the airstream (i.e. those shapes which are characterised by larger aerodynamic drag coefficients) will benefit less from smoothing of the underside than those with good aerodynamic shapes. This is shown diagrammatically in Fig. 3.17(a) which depicts the possible improvement of the drag coefficient achieved by smoothing the underside. The effect is expressed in terms of the ratio of the drag coefficient increment to the total drag coefficient of the car. The particular points of the curve represent values for typical cars.

Underside panel curvature

The flow underside is affected considerably by the shape of the panel, as expressed by its lengthwise and crosswise curvatures. The viscosity and consequent formation of the boundary layer leads to the formation of a choked airflow, i.e. to the blocking of the free passage of the airstream. It would therefore be beneficial to style the underside panel in such a way as to facilitate the airflow, making it more free. Increasing the average airflow velocity would lower the average pressure there, according to equation 2.1. Consequently this would decrease the aerodynamic lift of the vehicle, and reducing the momentum losses of the underside airstream would reduce the aerodynamic drag. The advantages of this will be explained in Chapter 7 which discusses car handling and stability problems.

The underside surface is usually symmetric with respect to the longitudinal axis of the car and can have a lengthwise and crosswise curvature, Fig. 3.17(b). We mentioned earlier that the underside flow is more restricted than that on the sides of the car. Thus the average air pressure along the sides is lower than on the underside and the airflow gradually turns towards the side edges of the underside panel. This cross-flow relieves the pressure in the underside region, and the pressure

---

Fig. 3.17. (a) Interrelation between total drag and the effectiveness of drag reduction by fitting a smooth underside panel; (b) lengthwise and sidewise curvatures of the underside panel; (c) affect on the wake of underside styling

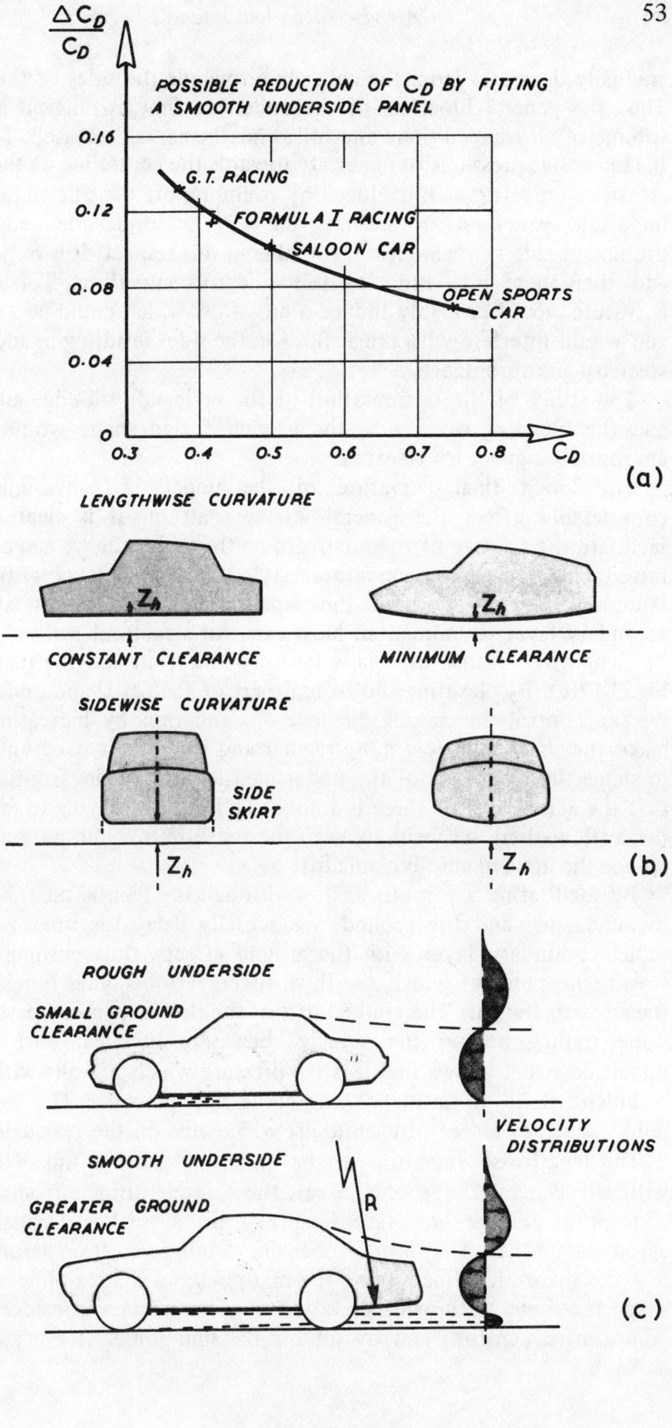

ΔC_D / C_D
POSSIBLE REDUCTION OF C_D BY FITTING A SMOOTH UNDERSIDE PANEL
0.16
0.12
0.08
0.04
0
G.T. RACING
FORMULA I RACING
SALOON CAR
OPEN SPORTS CAR
0.3
0.4
0.5
0.6
0.7
0.8
C_D
(a)
LENGTHWISE CURVATURE
Z_h
Z_h
CONSTANT CLEARANCE
MINIMUM CLEARANCE
SIDEWISE CURVATURE
SIDE SKIRT
Z_h
Z_h
(b)
ROUGH UNDERSIDE
SMALL GROUND CLEARANCE
VELOCITY DISTRIBUTIONS
SMOOTH UNDERSIDE
GREATER GROUND CLEARANCE
R
(c)

gradually decreases from the centreline towards the sides of the panel. Thus the general blocking of the underside flow is relieved and the volume of air trapped there and following the car is decreased. To allow the lower side pressures to penetrate towards the centreline of the panel, crosswise curvature is introduced by rounding off the side edges of the underside panel. As a general rule all flat undersides edged by prominent side skirts are less favourable in this respect. It is only fair to add that there is a limit to this sideways curvature. Too great a curvature would probably induce a crossflow which could be excessive and would interfere with the airflow on the sides resulting in additional vorticity and turbulence.

The study of the optimisation of the underside sill edge curvature and the blending of it with the general design shape would be an appropriate subject for research.

The longitudinal curvature of the underside panel may also considerably affect the general airflow pattern. It is clear that to facilitate the passage of the airstream on the underside we can consider introducing lengthwise curvature, styled to ease off conditions of boundary layer formation, the separation bubble effect and the secondary layer phenomenon. Most cars, for structural reasons, have a flat lengthwise frame, especially between the front and the rear axles, Fig. 3.17(c). By elevating the frontal part of the car shape underneath we can control the mixing distance d_m and thus by increasing it we make the underside flow more regular and smooth. It is advantageous to shape the front part of the underside, i.e. around the front axle, so that the actual airflow there is smooth and fast. By trying to facilitate the airflow there we help to keep the pressure low and consequently reduce the upward aerodynamic lift.

By facilitating a smooth airflow through the frontal part, between the underside and the ground, we actually delay the mixing of the vehicle boundary layer with the ground effects, thus pushing downstream the point at which the flow underneath becomes blocked and travels with the car. The choked part of the flow becomes a part of the wake trailing behind the vehicle. Elevating the rear part of the underside panel allows the negative pressure which prevails within the turbulent wake to influence the underside pressure. This helps to reduce the upward aerodynamic lift which acts on the rear axle [35].

The lengthwise curvature of the underside panel is linked closely with the general shape of the car; the side elevation introduces the concept of *camber line,* (see Chapter 4 for a further discussion on aerodynamic lift). The term camber line is borrowed from aeronautics, it is the theoretical line joining the mid-points of a streamline aerofoil shape. Looking at the side elevation of a car we may consider it as a comparative aerofoil, and by joining the mid-points at every station

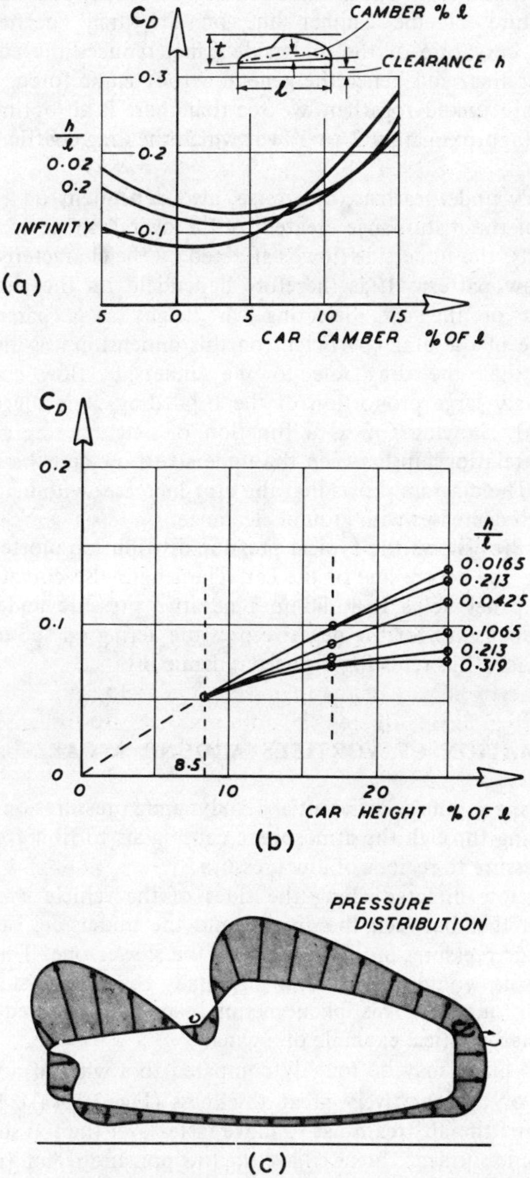

Fig. 3.18. (a) 'Camber' effect on the drag coefficient; (b) height effect on the drag characteristics; (c) comparison between pressure distribution on the upper and under surfaces

along its length draw a 'camber line'. Fig. 3.18(a) shows the effect of the curvature of the camber line on the drag coefficient. The lengthwise curvature of the underside panel reduces the curvature of the camber line, and hence the general aerodynamic force. These two problems are linked together. We see that there is an optimal camber curvature (approximately 2 to 5%) at which the drag coefficient C_D is a minimum.

The flow underneath is, of course, also dependent on the general intensity of the disturbance created by the overall shape of the car. In other words, the underside flow is affected by the characteristics of the general flow pattern. It is therefore dependent on the characteristic dimensions of the car. Selecting car height as a parameter the dependence of the drag coefficient on this dimension was investigated. Assuming that the drag due to the underside flow comprises a comparatively large proportion of the total drag, it is suggested that Fig. 3.18(b), showing C_D as a function of height t, represents to a degree the relationship between the underside flow and the rest of the car shape. The diagram shows that the drag increases with height, but at a rate which decreases with ground clearance.

Fig. 3.18(c) shows the typical pressure distribution plotted qualitatively along the centre line of the car. The lengthwise curvature of the underside panel helps to build-up a negative pressure under the car, which compensates for the negative pressure acting on the upper parts of the surface, thus reducing the aerodynamic lift.

3.6 FORMATION OF VORTICES AROUND A CAR

The car shape produces differential aerodynamic pressures on its surface when moving through the atmosphere causing air to flow from regions of high pressure to regions of low pressure.

The airflow directed along the sides of the vehicle is on average faster than that between the ground and the underside. In the faster airstream air pressures are lower than in the slower one. Thus air from the underside would tend to flow diagonally toward the sides, to join the stream outside. This phenomenon can be compared with the following aeronautical example of a wing.

The car shape may be roughly compared to a wing of a very short span and of comparatively great thickness (Fig. 3.19(a)). Due to its profile shape the airstream has to move faster over the top surface than the underside surface. Thus we have the low pressure region (marked by a negative sign) on the upper surface and the higher pressure region on the lower surface (marked by a positive sign). These two regions are separate but join along the edges, where air would naturally tend to flow from the underside towards the top. This cross-flow constitutes an aerodynamic interference between the two airstreams and often results in rotary air motion.

This rotary air motion is called a 'vortex' in aerodynamics. There are some differences between the theoretical assumptions of a vortex (the mathematical expression includes the idea of an infinite velocity at its centre) and its experimental form, where we observe the behaviour of the vortex motion as it occurs in nature. Such a vortex usually consists of a 'core', a cylindrical mass of air, rotating at a constant angular velocity like a solid mass, while the flow outside is induced to rotate at an angular velocity which is inversely proportional to the distance from the centre of the core. This phenomenon is well known and studied in aeronautics, and in theoretical aerodynamics, from where much of the terminology and knowledge is borrowed (see Fig. 3.11(a).

In aeronautics we can sometimes observe a vortex forming on the wing tips, or at the end of the flaps, especially when an aircraft is landing and its wings are at a high angle of incidence. This vortex motion originates about a car body, which has the same characteristics, and travels with it, but the two ends forming on the opposite sides of the vehicle extend back to the sides of the wake and continue downstream. The vortices forming part of the vortex train travelling with the car are called 'attached' vortices while those travelling behind the vehicle are referred to as 'trailing' vortices.

Driving behind a vehicle when it is snowing, or when there are loose fallen leaves on the road, or watching the swirl of exhaust fumes, we can easily see two distinct vortices on the outer sides of the car, i.e. masses of air on the sides of the wake rotating inwardly. The size and the intensity of these vortices is related to the difference between the actual aerodynamic pressure on the top and the underside of the car, on the car speed and on the general car shape. This pressure difference depends on the styling of the body, especially its 'camber line'. This pressure difference is also directly related to the aerodynamic lift component L_A—we say that the resulting vortices are 'induced by the aerodynamic lift'.

Aeronautical wing theory shows that a wing aerofoil induces a 'circulating' airflow around it in addition to that caused by the velocity of forward motion [36]. This circulating airflow which is superimposed on the velocity of motion is termed in aerodynamics *circulation.* This circulation in combination with the general flow pattern causes a pressure distribution resulting in aerodynamic lift.

It is known from the general laws of physics that vortices cannot terminate in space and are formed either in closed loops or terminate on a rigid boundary. It can, therefore, be deduced that the aerodynamic lift on a car is generated by the 'circulation', which is a vortex which

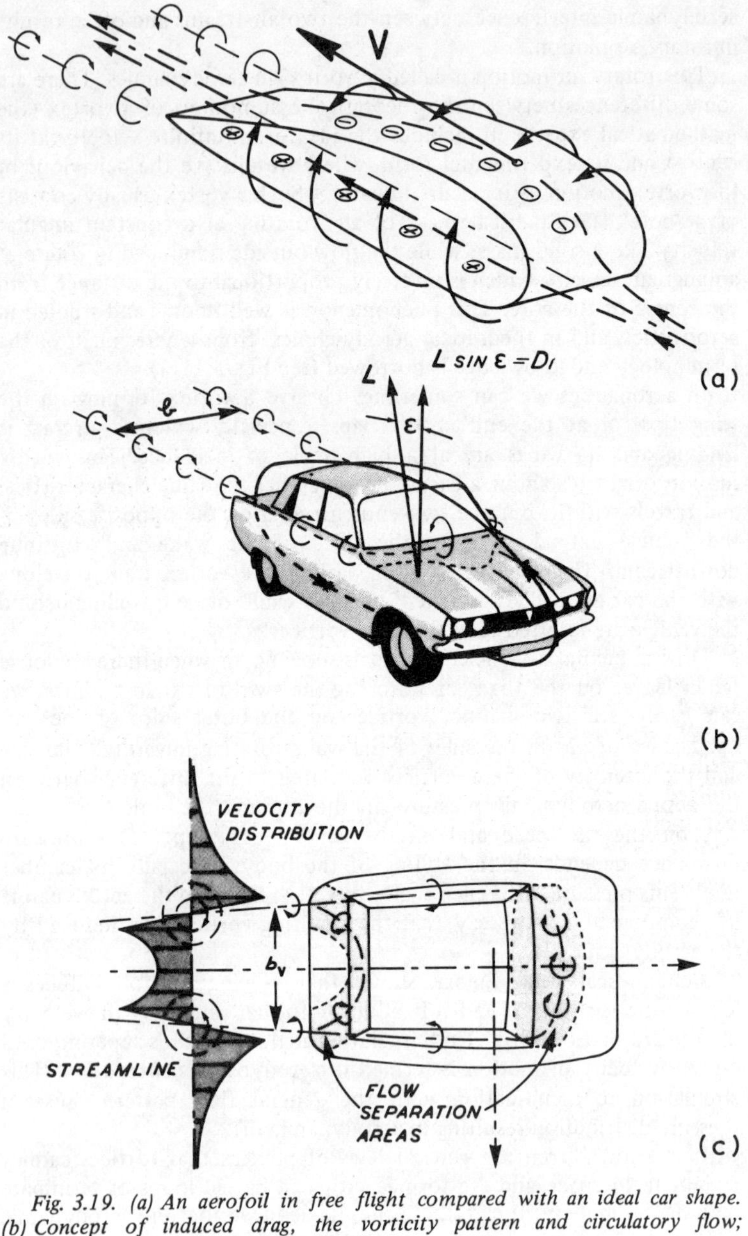

Fig. 3.19. (a) An aerofoil in free flight compared with an ideal car shape. (b) Concept of induced drag, the vorticity pattern and circulatory flow; (c) velocity distribution across trailing vortices

has its axis transverse to the direction of motion, the side vortices which are trailing behind are a continuation of the main vortex. Theoretically these vortices trail behind 'to infinity' but in reality they are gradually dissipated inside the rear turbulent wake, Fig. 3.19(b).

We know from aerodynamics that the relation between circulation and aerodynamic lift force is expressed by an explicit formula [26,28]. The effect of the side vortices on the general flow is to bend the central airflow downwards over the vehicle. This results in a rearwards deflection of the resultant aerodynamic force by a small angle ϵ, which is itself proportional to the circulation and hence to the lift coefficient. This deflection creates an additional aerodynamic force component D_i in the downwind direction which we call the 'lift induced drag' or possible better, the 'vortex induced drag'.

$$D_i = L \sin \epsilon = qSC_{Di} \tag{3.2}$$

Since the small angle ϵ is proportional to C_L, the value of the 'induced drag coefficient' C_{Di} is directly related to the aerodynamic lift coefficient

$$C_{Di} = \frac{L \sin \epsilon}{qS} = kC_L^2 \tag{3.3}$$

where k denotes a constant factor which aerodynamic theory shows to be proportional to the planform length to width ratio and dependent on planform shape. The analysis and derivation of this formula can be found in many textbooks on aerodynamics [21,26]. In the case of vehicle shapes we cannot directly apply some of the more simple but well developed aeronautical theoretical methods and we are therefore obliged to rely on experimental methods to estimate the factor k because of ground interference, underbody roughness and other effects.

For an average saloon car, the planform width to length ratio does not vary much from value of 0.37. Using this value, statistical analysis of observed overall lift and drag coefficients of a large number of cars has shown that, for an average saloon car, the value of the factor k is close to 0.41. Indeed it has been suggested [89] that the value of the induced drag coefficient of a saloon car can be estimated from the measured aerodynamic lift, using the formula

$$C_{Di} = 0.41\, C_L^2 \tag{3.4}$$

This formula should be applied with some caution to any individual car.

As we have already remarked, the styling of saloon cars has an important effect on the magnitude of the lift induced drag. It was found that both the front windscreen and the rear window may act as potential generators of vortices and we described (Fig. 3.2) the separation bubble which often forms at the corner between the bonnet and the lower part of the windscreen. The air movement within the bubble is circulatory. This rotary movement of air, which in due course becomes a vortex, extends towards the side edges of the windscreen. The flow along the sides acting on the windscreen edges, which are usually sloping backwards, may amplify the vortex. The vortex bends downstream and continues alongside, becoming a trailing vortex, Fig. 3.19(c). Thus the styling of the windscreen-bonnet configuration affects the position and intensity of the trailing vortices and the behaviour of the airstream over the upper parts of the car.

Similarly, the shape and the inclination of the rear windows may affect the secondary vortex which is generated either on the rear edges of the window or within the separation bubble forming at the lower edge over the car boot.

These secondary vortices may join together forming two pairs of vortices, an upper and a lower pair. These vortices are all of the same sign, i.e. they rotate clockwise. When this vortex system trails behind, leaving the car, the pairs interact and are drawn towards each other which results in the two pairs joining into a single pair. Wind tunnel investigations seem to indicate that the vortex pair near the ground is usually the weaker of the two and that the upper pair predominate. In practice it is difficult to observe the formation of the two distinct pairs of vortices, and only the one joined together is observed just behind the car. Sometimes the styling of the car is such that the streamlining of the upper parts weakens the upper pair of vortices and those nearer the ground (due to a combination of the flow along the sides and underneath the car) predominate. This vortex is usually more visible when driving behind a car.

The aerodynamic definition of 'circulation' is that it is the integral of velocity taken around a looped streamline.

In the case of a vortex the value of the circulation induced when the rotary motion produces a tangential velocity U is

$$H = 2\pi r\, U$$

where r is the radius of the given circular streamline within the vortex flow (see Fig. 3.20). The value of the velocity U can be measured in a wind tunnel behind the car scale model. From theory, the lift due to the main vortex, which is assumed to be of the same strength as the trailing vortex, is a direct function of the circulation H. Thus we can theoretically calculate the lift coefficient in terms of the circulation. Assuming that the centres of the two trailing vortices are approximately situated on the outside of the car, we can say that $2r = d$ when d is the width of the car, and the formula for the lift coefficient is

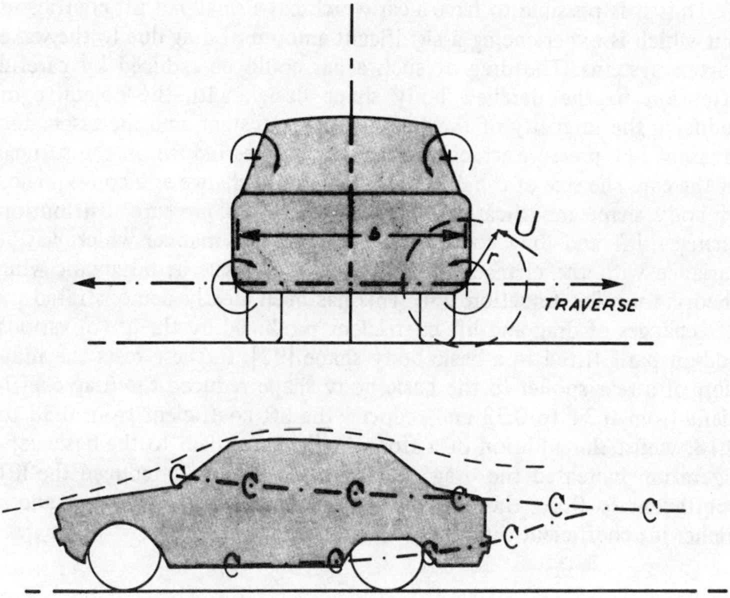

Fig. 3.20. The vortex pattern around a car

$$C_L = \frac{2\pi d^2 U}{SV} \tag{3.5}$$

Values of C_L obtained by this method do not correlate very well with direct measurements of the aerodynamic lift either on full-sized vehicles or in wind tunnel tests on scale models.

Since the actual mechanism of interference between the pairs of vortices is complex and involves energy losses, any measurements of velocity distribution in the turbulent wake field is very difficult and not often attempted. However, measurements which have been made [90,91] show clearly the existence of a primary vortex system and a number of secondary vortex systems. These vortex systems often originate from regions of the body where there is a significant change of surface slope or body cross-section area. Each vortex system will generate its contribution to the 'induced' drag coefficient irrespective of the sense of rotation within the vortices, or whether they generate positive or negative lift contributions. However, there is a direct relation between the net strength or the algebraic sum of the strengths of these trailing vortex systems and the aerodynamic lift generated by the shape of the car.

Thus it is possible to have a car which has a small net lift coefficient but which is experiencing a significant amount of drag due to the wake vortex systems. The drag of such a car could be reduced by careful attention to the detailed body shape design with the objective of reducing the intensity of the trailing vortex system, and the associated areas of low pressure which the trailing vortices induce on the surface of the car. The size of the lift coefficient could change as a consequence of body shape modifications and the associated pressure distribution changes, lift and drag coefficients changing in a manner which was at variance with the changes predicted by the classic aerodynamic wing theory formula, Equation 3.4. This has been clearly demonstrated by the changes of drag and lift coefficient produced by the use of various add-on parts fitted to a basic body shape [92]. In these tests the addition of a rear spoiler to the basic body shape reduced the drag coefficient from 0.34 to 0.33 and reduced the lift coefficient from 0.38 to 0.18, whilst the addition of a front spoiler or airdam to the basic configuration increased the drag coefficient to 0.38 and reduced the lift coefficient to 0.29; the original configuration had a lower drag and a higher lift coefficient than the final configuration.

3.7 STYLING OF STREAMLINED BACK

When discussing the general airflow pattern over the car, and 'bobtailing', we remarked that in order to reduce the intensity of turbulence and the size of the wake behind the car it is better to style the rear surface so that it slopes gently backwards to avoid premature flow separation. At a line at which the separation would be inevitable due to viscous friction the shape is cut, or 'bobtailed', thus forming the rear base. This approach is generally adopted by many sports and fast car stylists–the so-called 'fastback'. Fig. 3.21 shows a scale model of a saloon car with a fastback rear. Using one of the flow visualisation methods, which involves covering the surface of the model with special liquid and exposing it to the airstream, we are able to photograph the impact of the flow pattern on the model surface. On the side edges at the rear we can see the impact of a vortex probably induced by the edges of the rear window. From the impact pattern and the configuration of the streamlines we can deduce that the vortex is rotating in an anticlockwise direction. Considered three-dimensionally, this would induce the airflow moving along the upper parts of the car to be bent downward. Thus the presence of trailing vortices causes an induced downwash airstream resulting in the inclination of the streamlines downwards behind the main car cabin roof. The effect of the trailing vortices is therefore beneficial, in a way, in reducing the

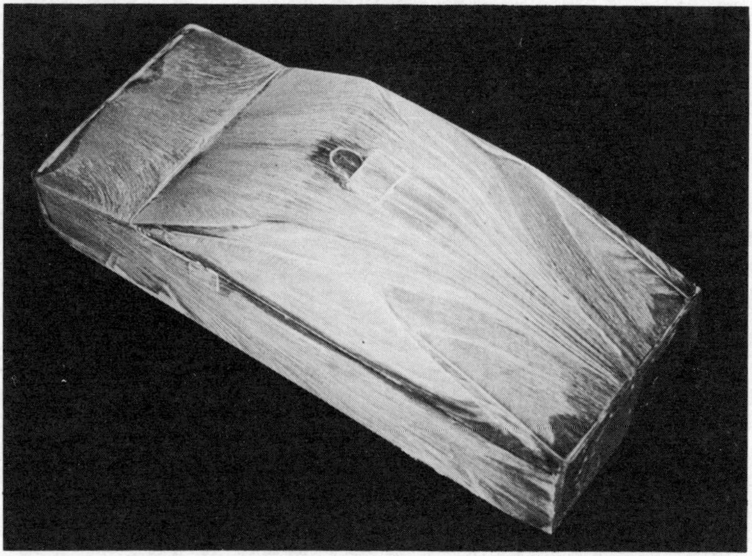

Fig. 3.21. Airflow visualisation (vortex traces) on scale model of fastback car

possibility of premature flow separation on the rear surface. This is quite effective, more in fact than any of the artificial devices used to keep the airstream steady. Thus a pair of well-developed vortices helps to keep the airstream smooth, and faster, making fastback shapes effective.

Whilst the fastback shape is useful for the 2+2 coupe or sports car shapes it has been adapted by many volume manufacturers into the hatchback by increasing the angle of deflection from the roof line to the rear window. Flow investigations [92-94] show that as this angle changes there are changes in the flow pattern over the rear of the body, with consequent changes in the drag and lift coefficients.

As the deflection angle is increased out of the fastback range of less than 20° and into the hatchback range, for angles up to about 30° there is a steady increase in the drag of the car and in the strength of the trailing vortices which are situated just inboard and to the rear of the D-posts as shown in Fig. 3.22(a). It is their downwash which ensures that the flow passing over the rear edge line of the roof remains attached to the inclined fastback surface, as described above. Despite the reduced area of the wake that accompanies this flow as it remains attached towards the rear end of the car, the drag increases steadily to a maximum due to the high suctions which the vortices induce on the inclined rearwards-facing surfaces over which they are located. Because

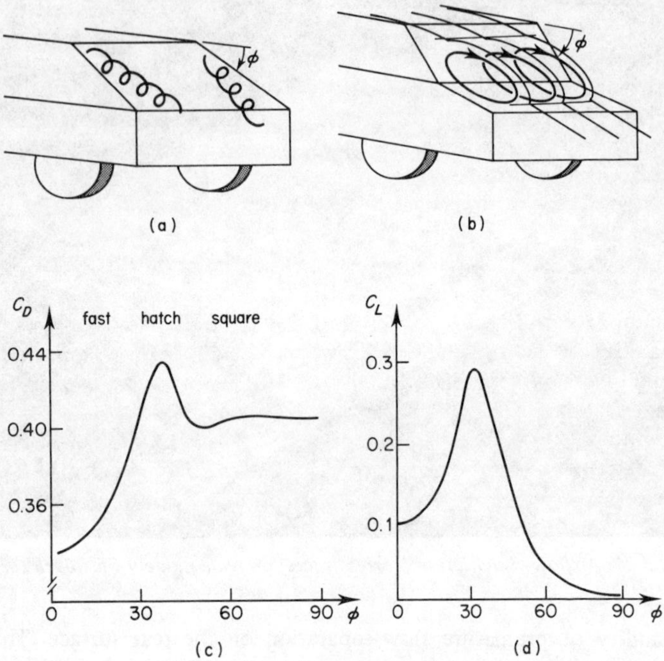

Fig. 3.22. (a) Vortex structure over fastback and (b) hatchback rear end; (c) dependence of drag coefficient and (d) lift coefficient at rear axle on rear surface declination angle.

of the relatively large area of this low-inclination surface over which these low pressures are generated, there is a significant lift coefficient, which may reach 0.25, acting at the rear axle of a fastback car.

As the rear surface deflection angle increases slightly more, it enters a critical region in which the flow is not fixed, the trailing vortices separate intermittently from the fastback surface and the general flow separates intermittently from the rear edge line of the roof rather than from the rear end of the fastback surface. This change in flow pattern occurs as the deflection angle changes through 6° to 8° and is accompanied by a rapid reduction in the average drag coefficient and average lift coefficient acting over the rear wheels, as shown in Fig. 3.22(c) and (d). The values of the critical angle range depend upon a number of factors which include the other geometric properties of the body shape, the pitch and yaw angle of the body and the turbulence in the approaching airflow.

For further increase in the rear surface deflection angle through 50°, the normally accepted upper limit for a hatchback, and on into the squareback configuration with deflection angles up to 90°, the drag coefficient is almost constant. There is little change in the pattern of the flow over the rear of the body with the flow separating from the rear ridge line of the roof and giving almost constant pressure over the rearward facing surfaces. Hence in addition to the almost constant drag coefficient, there is also very little lift coefficient generated over the rear wheels.

The notchback configuration is essentially a hatchback with a rearwards extension from below the lower edge of the rear window to accommodate a boot. A slightly higher flow deflection angle can be tolerated with a notchback than with a hatchback rear end, and a different vortex system may develop with a transverse vortex being present in the locally-separated flow region or 'bubble' set around the rear window and boot lid junction as shown in Fig. 3.22(b). The flow may reattach towards the rear of the boot lid, giving a region of higher pressure which will have little or no effect on the drag coefficient due to the inclination of the surface, but which will reduce the rear end lift coefficient.

In almost all of these rear-end configurations, the shape of the rearmost upper edge from which the flow finally separates to form the wake has a significant influence on the drag of the car. If the edge is radiused, the flow will tend to follow the curved surface for some distance and generate a suction before separating, which will contribute to the drag. This may be strongly affected by cross-wind conditions and other factors which can influence the location of the point of flow separation and the steadiness or stability of the flow pattern. If the rearmost upper edge is square or even slightly upturned to form a rear spoiler, it will eliminate the local suction peak and fix the location of flow separation, thereby giving a lower drag coefficient and a more stable flow pattern over the rear end.

3.8 INTERNAL AIRFLOW

An analysis of car aerodynamics cannot avoid the problem of the internal airflow. In fact the car engine, and sometimes the brakes, requires a quite considerable mass of air to be drawn through the inside of the car. This mass of air is taken from the outside flow and induced to pass through a system of internal ducts and cavities. This airflow is either assisted by a fan operating within the internal duct, or it utilises the external pressure differences between the inlet and the outlet of the internal duct. The internal airflow has a two-fold effect on the external airflow pattern. First, by depriving the external flow it lowers the overall pressure, and second, the mass taken from outside and induced to flow through an internal duct, loses momentum through friction turbulence and leakage. The first effect on the external airflow is, theoretically, beneficial, i.e. bleeding a mass of air into the internal duct system reduces the value of the resultant aerodynamic force. On the other hand the internal flow introduces additional drag. Balancing one against the other, the losses are always greater than the gains and we have in effect the problem of an 'internal aerodynamic drag'.

Discussing this internal aerodynamic drag with stylists and engineers, one feels that the problem of the efficiency of the cooling, or ventilation, is the main concern and that the aspect of the aerodynamic effect on the car performance is not fully appreciated.

The internal airflow constitutes a special design and styling problem. In practice the internal airflow is concerned only with the engine cooling system, and in some cases the brake cooling system; interior ventilation involves a comparatively small mass of air and consequently any momentum losses are small and can be neglected from the aerodynamic airflow pattern point of view. Engine and brake cooling systems require comparatively large airflows, depending on the size of the engine and the speed of the car. The aerodynamic problem of internal ventilation is important in itself and requires special attention but it is primarily a problem of the slow movement of air and does not affect the general external airflow pattern.

Problems of ventilation are only related to the overall aerodynamic styling in the practical choice of inlets and outlets of the ventilation system so as to utilise the existing pressure distribution and pressure gradient as a means of flow induction.

The first step in the investigation of the cooling air flow system is to estimate the quantity of the excess heat to be dissipated for a given vehicle speed. For a typical heat exchanger, i.e. the radiator, this quantity can be calculated from the formula

$$Q = K_1 A(T_w - T_A){V_2}^{0.8}$$

where A is the area of the radiator and K_1 is a constant factor dependent on the type of radiator [37-41]. If K_1 is known, then the temperature T_W of the walls of the radiator and the air temperature T_A, together with the local airflow velocity on the entry to the radiator V_2, we may estimate the excess heat dissipated from the engine. Car designers use larger sizes of radiator than necessary, thus making certain of the spare heat dissipation capacity. The size of the radiator and the velocity of cooling air at the entry to it when multiplied by the air density would give the rate of the air mass flow required:

$$dm = \rho A V_2 dt.$$

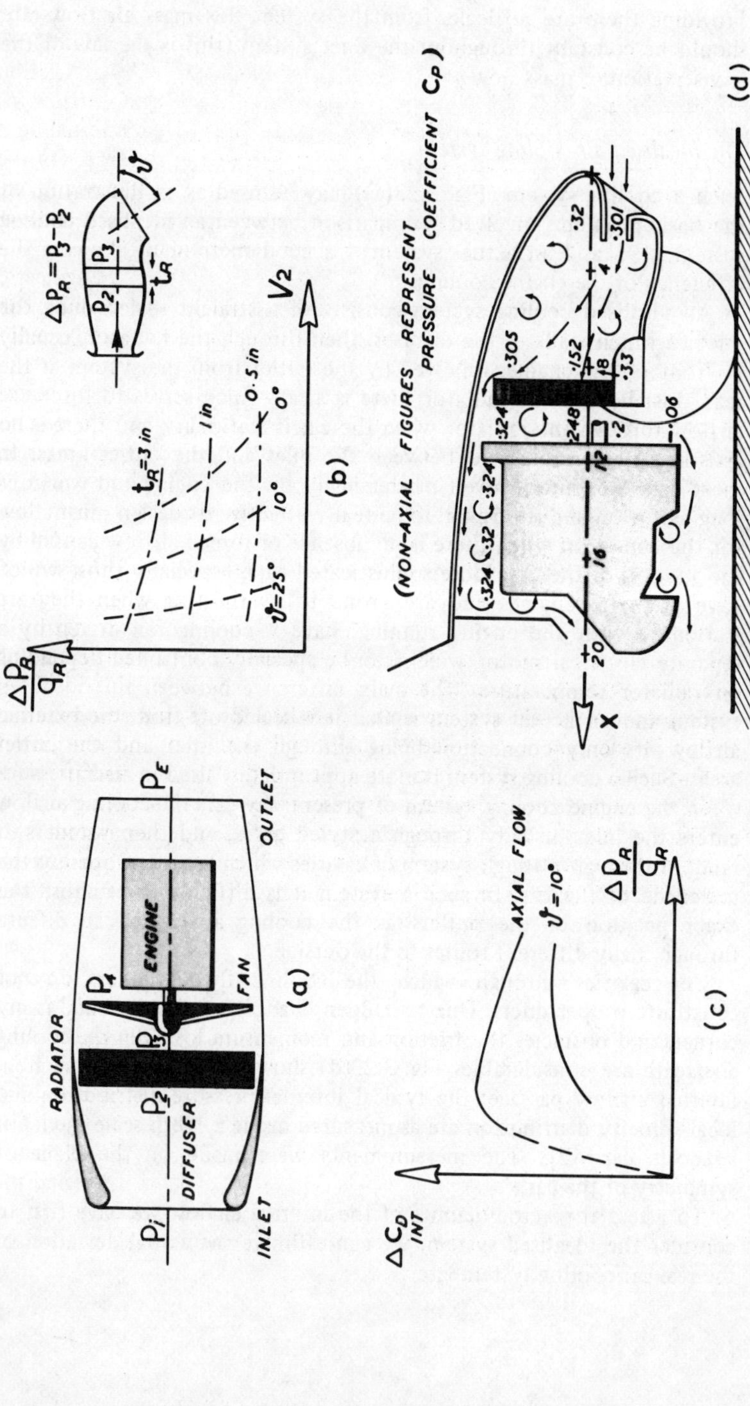

Fig. 3.23. (a) Ideal cooling-flow system; (b) effect of radiator thickness and inclination of the outlet; (c) effect of pressure differences across the radiator on drag; (d) airflow pattern inside a typical engine compartment

Providing there are no leaks from the system this mass air flow rate should be constant throughout the duct system (this is the law of the 'conservation of mass flow').

The idealised air cooling system

Such a cooling system (Fig. 3.23(a)) may be used as an illustration of the basic problems involved. Comparison between an idealised cooling system [19] and an actual system is a good method of showing the efficiency of the engine cooling.

An idealised cooling system consists of a straight sealed duct, the inlet to which leads to the diffuser, then through the radiator, usually in front of the engine, followed by the outlet from the system at the rear. Just behind the radiator there is a fan which serves to force the airflow through the radiator when the car is stationary and there is no assisting pressure gradient between the inlet and the outlet areas. In most cars the fan is driven mechanically by the engine and works as long as the engine is turning; to be really effective its design must allow for the condition when there is an absence of forced airflow caused by the motion of the car. More sophisticated cars, especially those which have powerful engines and are prone to overheating when they are stationary with the engine running, have a cooling fan driven by a separate electrical motor which can be specially controlled depending on radiator temperature. The main difference between this idealised system and an actual system is that it is sealed off from the external airflow–its only connection being through the inlet and the outlet areas. Such a cooling system is more appropriately used in aircraft.

In the engine cooling system of present day cars the cooling airflow enters the inlet, usually through a styled grille, and then wanders at random through a rough system of cavities which are often open to the underside of the car. In such a system it is difficult to establish the exact position of the outlets as the cooling airstream can diffuse through many different routes to the outside.

The cavities through which the cooling flow wanders do not constitute proper ducts. Due to sudden changes of direction and many corners and obstacles the friction and momentum losses in the cooling airstream are considerable. Fig. 3.23(d) shows an example of such an internal airflow passage; the typical internal pressure distribution and local velocity distribution are as measured inside a 1/8th scale model of a sports car [42]. The measurements were made on the plane of symmetry of the car.

To assess the aerodynamics of the internal airflow we have first to consider the idealised system. We can estimate the actual deviation of the real car cooling system later.

The inlet has an area A_i (Fig. 3.23(a)) and the air-mass flow rate through it is dm/dt kg/s. This flow enters the part of the duct where should theoretically be shaped so as to induce air to slow down to the optimum speed required for efficient cooling through the radiator. This part is called the *diffuser.* Something resembling a diffuser can be found in most cars. Usually we find, behind a frontal grille, a short cowling leading to the radiator which could be considered as some sort of diffuser.

The aerodynamic efficiency of the radiator, denoted as η_R is defined as the amount of the heat dissipated per unit force of the aerodynamic drag. It can be expressed as follows:

$$\eta_R = \frac{Q}{D_R} = \frac{2K(T_w - T_A)}{\rho V_2^{1.2} C_{Dr}} \tag{3.6}$$

The formula shows that the efficiency of the radiator increases as the airflow velocity, V_2, through the radiator decreases.

Many car stylists seem to misunderstand this problem and they provide a short duct in front of the radiator with a gradually decreasing instead of increasing cross-sectional area. This has the adverse effect of speeding-up the airflow. Most car designers do not provide a diffuser and the passage between the inlet grille and the radiator is left as an irregular cavity without distinct walls.

The idea behind the diffuser is to regulate the cooling airflow to enable it to enter the radiator at just the right speed for the optimal exchange of heat to be transferred between the radiator walls and the air. For maximum efficiency the airflow must be sufficiently turbulent for the air particles to come into contact with the walls and to have sufficient time to collect as much heat from them as possible. To do this a slow airflow is required, as equation 3.6 shows. This part of the duct should therefore be shaped so that the duct cross-sectional area increases gradually. The front grille creates the required conditions of turbulence. It is known from experimental aerodynamics that the walls of the diffuser duct should not decline more than 7° to the centre line of the duct at any place otherwise the flow would tend to separate from the walls and local eddies would form. Such eddies not only make the airflow more difficult to control, adding to the internal flow drag, but would allow some parts of the radiator to become hotter than others. Nevertheless the airflow through even a well designed entry diffuser involves a certain amount of momentum loss. This is due to the wall friction and small irregularities of manufacture. We usually measure this momentum as a drop in the total pressure H (see equation 2.1).

Many car designers seem to be concerned only with adequate cooling of the engine or brakes. When this is proved to be sufficient the problem of the internal airflow is usually neglected. In fact, as the air mass concerned is considerable and its momentum losses comparatively great, the total aerodynamic drag involved plays a marked role in the total car drag.

The actual aerodynamic drag due to the internal flow can be expressed by the total loss of momentum, or work done, due to the passage of the cooling air through the system. The loss of momentum is expressed by the product of the rate of mass flow and the change in velocity between the inlet and outlet of the cooling duct, i.e.

$$(V_i - V_e)\frac{dm}{dt}$$

where V_i and V_e are the inlet and outlet velocities respectively. The work required to overcome friction and internal obstructions, such as the radiator, can be expressed in terms of the pressure drop multiplied by the exit area, i.e.

$$(p_i - p_e)A_e$$

Thus the aerodynamic drag of the internal airflow is expressed as

$$D_{int} = (V_i - V_e)\frac{dm}{dt} + (p_i - p_e)A_e \tag{3.7}$$

And the internal drag coefficient is

$$C_{D_{int}} = \frac{(V_i - V_e)}{qS}\frac{dm}{dt} + \frac{(p_i - p_e)A_e}{qS} \tag{3.8}$$

These equations show how important is the selection of the outlet area. The drag increment due to the internal airflow depends on the mass flow involved, on the difference between the air velocity at the inlet and the outlet, as well as the difference between the pressures at the inlet and the outlet. Finally it is dependent on the ratio of the inlet area to the total cross-sectional area of the vehicle S, and the ratio of the outlet to the total cross-sectional areas.

We have to remember that these calculations assume that the air passes from the inlet to the outlet through a sealed duct. In reality airflows are rather different to those shown in Fig. 3.23(a). In practice, air escapes mainly around the engine compartment, through apertures at the underside or through the bonnet lid around the edges (usually the local air pressure over the bonnet is low because the airstream velocity is high and this creates a strong suction). Many vehicles exhaust the cooling airflow directly downwards from the engine cavity which is open to the ground. The flow on the underside and just downstream of the front wheel axle is quickly slowed down due to the obstruction further back but it still flows freely and the comparatively large area around the engine crankcase from which the slow flow emerges causes a lot of confused eddies to form. Thus the total loss of momentum from the underside and internal flow is considerable.

Without exception the inlet to the cooling system is placed at the front of the car. This is the region of highest dynamic pressure, i.e. the region where the kinetic energy of the airflow is changed into static pressure (see equation 2.1). In front of the radiator grille pressures are, therefore, higher than atmospheric. At other parts of the surface, the local pressure depends on the local velocity, the higher the velocity the lower the pressure. It is therefore important to position the outlet for the cooling flow where the actual pressure is lower than atmospheric. Such a region is either in the wake, at the rear of the car, or at the sides where the airflow velocity is greatest, and consequently, where the suction assists the internal flow to discharge to the atmosphere.

The mechanical energy taken from the engine by the 'vee' pulley and belt driving the fan is imparted to the internal flow. This energises the flow and thus counteracts the loss of momentum suffered during its passage through the ducts. It is additional energy taken from the power source driving the vehicle and is not therefore directly involved with the estimation of aerodynamic drag.

There is always an advantage in reducing the internal flow drag if possible. We may, in general, conclude that the following measures would cause a reduction in this drag:

(1) Placing the inlet area at the front of the vehicle, i.e. in the region of the highest aerodynamic pressure.

(2) Placing the outlet in the region of low pressure and making the area of the outlet adjustable.

(3) Sealing the whole cooling system, if possible, to avoid any leaks to the outside airflow.

(4) Arranging the diffuser to slow down the internal flow to the velocity required for maximum radiator efficiency.

(5) Avoiding sudden changes in the internal flow direction and speed by arranging for the internal cross-sectional areas to change gradually and, if possible, in a straight line.

The amount to which the cooling drag can be minimised depends on the type of the vehicle, its external shape, the size of the engine, and on the arrangement of the cooling system and its components. Fig. 3.23(b) shows the importance of the change of direction of the flow of the cooling stream; the axial airflow through the system differs greatly if the flow is deflected by 10°.

Experimental Simulation of the Internal Cooling Flow

Designers of new types of cars always face the problem of preparing a preliminary estimate of the aerodynamic characteristics before the prototype is built, which can be very expensive. It is a problem of optimising many different and interrelated requirements. To do this, especially from the aerodynamic point of view, the designer, after choosing the shape of the vehicle, must find its basic aerodynamic lift and drag characteristics. This is usually done on scale models in wind tunnel tests. Accepting that the problem of the internal flow is aerodynamically important, how it is estimated at this stage?

The exact simulation of the internal flow in scale models is difficult in practice. The car industry uses scale models as large as the available wind tunnel permits–1/4 to 1/5th scale models are preferred (many university laboratories having smaller wind tunnels use much smaller models). It is possible to build internal cooling ducts into such models but to simulate actual operating conditions the simulation should include the radiator with its heat exchange effect, i.e. the hot water circuit. Most wind tunnel experiments neglect the simulation of the internal flow entirely, as being too expensive and thus the internal flow pattern around these models differs considerably from the actual pattern found when the car is on the road. Figure 3.24(a) shows the experimental difference between the internal cooling drag (expressed by the ratio $\Delta C_D/C_D$) of the system with and without heating of the cooling airflow by the radiator in model tests. Experimental data for the aerodynamic drag and lift coefficient, measured on a model of an average saloon car with its cooling airflow blocked and unblocked, indicates that unblocking the cooling air passages could result in an increase in drag coefficient of up to 0.10 and a reduction in lift coefficient of up to 0.12, but these values depend on the amount of flow passing into the cooling system.

The cooling drag is dependent on the mass flow through the cooling system and its change of streamwise velocity. This is influenced by the relative pressure drop across the radiator matrix which is, in turn, dependent on the mean speed of the airflow through the matrix and also on the temperature of this air. Thus for a typical matrix we find that a 10% increase in the pressure drop coefficient results from the change from 'cold' to 'hot' operation at all speeds whilst the pressure drop coefficient may change from 12 to 6 as the wind speed at its face increases from 10 ft/s (3 m/s) to 40 ft/s (12 m/s). Radiator matrices are often simulated in wind tunnel models by the use of one or two sheets of woven wire gauze which can be selected to generate the required pressure drop characteristic representative of typical high or low speed driving conditions.

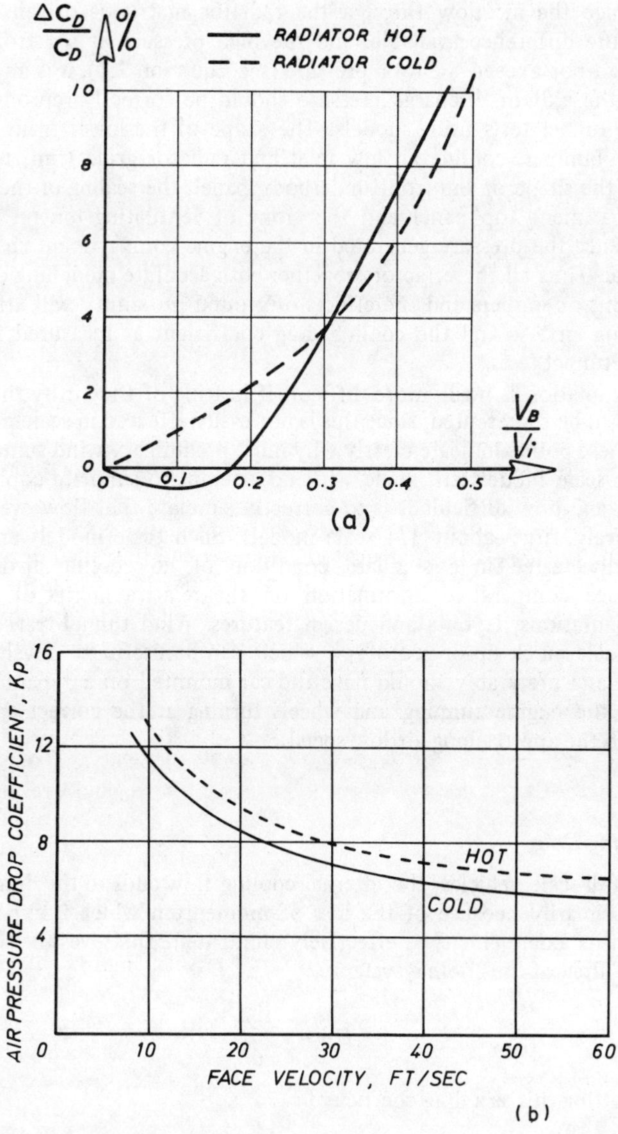

Fig. 3.24. (a) Radiator blockage and matrix temperature effect on vehicle aerodynamic drag; (b) temperature and speed effect on matrix pressure loss coefficient

Since the air flow through the radiator matrix is driven by the pressure difference available, and the high pressure at the front of a car can not exceed the total pressure (see Equation 2.1), it is important that the exit or discharge pressure should be correctly reproduced in wind tunnel tests using models. The shape of the lower front panels, front bumper, cooling airflow inlet and radiator grille trim, together with the shape of the front underbody panel, the sealing of the engine compartment top panel and the siting of ventilating louvres will all influence the pressure generated in the engine compartment space of a model. Thus all these factors together with accurate modelling of front airdams or spoilers and the effects of ground proximity will affect the cooling airflow and the cooling drag coefficient as measured in such wind tunnel tests.

Simulation is made more difficult if heating of the air by the radiator is to be represented, since this is not easily effected in scale models.

These points indicate clearly why most preliminary wind tunnel tests using scale models are made with no attempt to simulate cooling airflow, and how difficult it is to correctly simulate that flow even using relatively large–about 1/4 scale models. Such scale models are most usefully tested in a standard condition of no cooling airflow, to produce comparative information on the relative merits of various configurations, layouts and design features. Wind tunnel tests using a full scale mock up or prototype vehicle can be useful for development work and preferably should have the car mounted on a dynamometer, with the engine running and wheels turning at the correct speed to match the approaching airflow speed.

Cooling Drag

For full-scale vehicles, the internal cooling flow adds to the drag of the car primarily because of the loss of momentum which it experiences when its axial velocity is effectively eliminated. This gives the so-called 'sink' drag of the cooling system

$$D_R = V\frac{\mathrm{d}m}{\mathrm{d}t}$$

or, setting this as a drag coefficient

$$C_{DR} = \frac{D_R}{qS} = \frac{VA_R\rho V_2}{qS} = 2\frac{A_R}{S}\frac{V_2}{V} \tag{3.9}$$

where V_2 is the average speed through the inlet to the radiator having area A_R.

Measurements [92] on a large number of cars for the velocity and area ratios which appear in Equation 3.9 show that

$$0.02 \leqslant C_{DR} \leqslant 0.06$$

Wind tunnel measurements of the change in drag coefficient between tests made with the radiator flow open and the radiator blanked off showed that eliminating the cooling air flow reduced the measured drag coefficients of the various vehicles by amounts varying between 0.01 and 0.06, with an average of 0.03, This value is closely supported by the results of similar tests made at MIRA [93]. These results indicate that a properly designed cooling air duct system can achieve an increment of drag coefficient due to cooling flow as low as 0.01 or about 2.5% of the total drag of a car. On the other hand, the drag increase associated with the cooling airflow can reach 10% or more of the overall drag if careful attention is not paid to this matter.

In estimating the lowest possible drag coefficient for a cooling system, care should be taken not to be misled by changes in the gross drag of a vehicle as determined during wind tunnel tests. Blocking off the internal flow may cause the air, which would have passed through the engine compartment, to spill out over the bonnet and, by causing additional separated flow regions over the leading edge of the bonnet or side wings, to generate an increase in the drag of the external body. This is particularly likely to occur with cars which have only marginal rounding of the bonnet leading edge and then the cooling system drag coefficient will appear to be significantly smaller than its true value. A car which has been carefully designed for minimum drag throughout will, most likely, have more than marginally rounded leading edges and then errors in estimating the cooling flow and cooling system drag will be relatively more significant.

Analysis [95] of the internal flow requirements of vehicles has indicated that the minimum value of cooling drag coefficient that can be obtained whilst meeting the engine cooling-air requirements is about 0.014, and similar values have been achieved [96,97] during the development of cars with low drag.

More and more saloon car designers have recognised the wastefulness of having a continually running fan, driven by a belt from the in-line engine, mounted at the rear of the radiator matrix. This arrangement has often been replaced by a fan driven by an electric motor which only operates on demand, as sensed by a thermostat mounted on the engine or radiator, and which freewheels or windmills when not being driven. This eliminates the waste of the power absorbed by continually belt-driving a fan to force air through the radiator matrix, and thereby potentially overcooling the system at most road speeds above minimum.

Also, it has encouraged the wider use of transversely mounted front engines with front wheel drive.

At normal ambient air temperatures and engine operating speeds, the ram-induced airflow through the radiator matrix should be more than sufficient to provide adequate cooling if the system is properly designed. However, there will be a penalty on vehicle drag and performance if the ram effect is dissipated by blocking the approach to the radiator with a sharp-edged grille or a bumper bar. This penalty will be increased if the airflow is then allowed to spill around the radiator matrix rather than being made to pass through it, and if a clear passage is not provided for the heated air to escape from the engine compartment.

A cooling system is normally required to dissipate an amount of power which is between 60% and 70% of the gross brake power being generated by the engine at any particular engine speed. A typical radiator matrix can dissipate up to about 2000 BTU/min/ft^2 or 46 h.p./ft^2, but this depends on the detailed geometry of the layout and arrangement of the water tubes and of the fins. These factors in turn affect the air pressure drop coefficient, which typically lies between 5 and 10 (see Fig. 3.24(b)), and hence the speed of the airflow through the matrix for a given applied pressure difference. A matrix with a lower pressure coefficient will generally have a higher airflow rate and larger heat dissipation rate at a given road speed and engine power condition, and hence can be made smaller in size. This can then lead to a lower profile for the engine compartment and a consequent reduction in the overall drag of the body shape. The optimising of airflow pressure drop and heat transfer characteristics, together with material and manufacturing costs is still a matter of active investigation by many manufacturers.

In a specific problem investigated at MIRA [98], it was found that the radiator matrix area could be reduced to 75% of the area of the original production version if the matrix design was not changed, or to 45% of the original area if a redesigned but commercially available matrix were to be used. The power absorbed by the belt-driven fan was reduced from 1.0 h.p. to 0.3 h.p. at a road speed of 70 m.p.h. (110 km/hr) by redesigning the blades. The drag coefficient of the original car with cooling air inlet blanked was 0.469 whilst with the engine running and the cooling air inlet open it was 0.490. After modifying the fan and installing light ducting over the fan and matrix, the drag coefficient with engine running was reduced to 0.479 and, at 70 m.p.h. (110 km/hr) road speed this change in drag coefficient represented a change in the power absorbed by the cooling drag from 1.1 h.p. to 0.6 h.p. Hence there was a total saving of 1.2 h.p. due to the changes in the cooling system, with about 60 h.p. maximum available at the wheels.

The needs of competition cars impose additional demands on the designer to supply cooling and ventilating airflows. For instance in the

GTP class of cars in the Le Mans 24 hour event, vehicles may have a drag coefficient as low as 0.27 whilst being powered with a 550 h.p. turbocharged engine. At a top road speed of about 220 m.p.h. (360 km/hr) the engine water cooling radiator would pass about 210 ft^3/s (6 m^3/s) of cooling air but additionally the engine oil cooling radiator would pass about 105 ft^3/s (3 m^3/s), gear box oil cooling would require 35 ft^3/s (1 m^3/s) and the turbocharger intercooler would take 28 ft^3/s (0.8 m^3/s). Front and rear brakes, engine compartment and the passenger compartment would all require additional ventilating air.

Ventilation Air Drag

The flow of air into a car for ventilation purposes also adds to the drag of the vehicle due to the change of momentum of this air. The ventilation drag

$$D_V = \rho Q V$$

where Q is the volume flow rate of ventilating air passing through the passenger compartment. Ventilating airflow rates are generally given in terms of the number n of air changes in unit time which are required to give or maintain a required environmental condition. This is clearly influenced by many factors which include the number of occupants of the vehicle, whether or not they are smoking, as well as air temperature and humidity. The resulting drag coefficient is

$$C_{DV} = 2\frac{n\forall}{SV} \tag{3.10}$$

where $\forall$ is the internal volume of the passenger compartment. For typical saloon cars, the ratio of internal volume to reference area is about 10 ft (3 m), and it is unlikely that more than about 25 air changes per hour will occur. Hence at a road speed of 50 m.p.h. (80 km/hr) the ventilation drag coefficient would be about 0.0019 whilst at 70 m.p.h. (110 km/hr) it would be reduced to about 0.0013 (but giving a higher drag force); these values are close to the limits of current experimental accuracy.

It is widely observed that opening the windows of a saloon car results in the establishment of fluctuating or buffeting airflow into and out of the body compartment. Whilst it would be expected that there would be an associated increase in the momentum or 'sink' drag, it has not yet been measured in wind tunnel tests, nor has it been possible to determine any general indication of significant drag changes occurring by examination of the high speed performance of vehicles.

There is, however, evidence of a change of external pressure distribution over the rear doors of a light delivery van due to the opening of the front windows. During road tests these pressure coefficients were found to change from around −0.1 with the windows closed, to about +0.2 with the windows open and this gave a large change in the drag of the body. Thus there is scope for further work in this area.

In conclusion we see that at present the proportion of the total aerodynamic drag of an average saloon car which is due to the internal flow drag may be in excess of 12%. By redesigning the internal flow ducts and cooling system, it is currently possible to reduce the drag coefficient due to the internal flow to about 0.02 or say 5% of the total drag. There appears to be some scope for further reduction in this drag, which will become increasingly important as the overall drag coefficients of production cars steadily reduce to or below 0.30.

The main characteristic of the internal cooling air flow is that the air is taken from the main flow at a point on the external surface of the vehicle and, after passing through some (more or less defined) internal ducts, it is discharged back to the external flow at another point and with much lower momentum. A preliminary aerodynamic study of a new car type should therefore include a practical study of the most beneficial points on the external surface to place the inlet and outlets of such a cooling system. This can be done by careful plotting of the external pressure distribution. The designer would try to improve the internal airflow by arranging the inlet at that part of the external surface where the local air pressure is greatest, and the outlet where it is sufficiently low. This explains the fact that almost all car designers place the inlet in the front of the vehicle, but the positioning of the outlet is usually haphazard. In many cars, as we mentioned earlier, the cooling air is dissipated into the flow underneath the body, through the irregular gaps between the crankcase and the body structure. To decrease the internal flow aerodynamic drag and power loss it would be better to seal the underneath of the engine compartment and to arrange for a smooth and efficient duct with an outlet situated at the rear of the vehicle in the wake region where the pressure is low or at the rear end of the engine compartment.

Chapter 4

AERODYNAMIC LIFT

The modern trend towards better and faster roads, and consequently the development of cars able to cruise constantly at high speeds, has accentuated the problem of aerodynamic lift.

The aerodynamic lift acts, by definition, in the direction normal to the direction of motion, thus in the case of a car it acts normally to the ground. It is accepted as a universal sign convention that the lift is positive when it acts upwards, i.e. away from the ground, and is negative when it presses the vehicle to the ground.

In aeronautics we create special shapes to generate a positive lift force while in the case of cars we are interested in reducing or even reversing its sign. In discussing the general airflow pattern around the car we stated that car motion in close proximity to the ground always tends to generate a positive lift, and that this phenomenon is very sensitive to the ground clearance. Positive lift reduces the tyre load on the road and substantially lessens the grip. A more detailed discussion of this subject will be found in the chapters dealing with car stability.

The importance of the action of the aerodynamic lift on the road holding characteristics can be illustrated by considering the positive lift force acting on the front axle of a typical saloon car travelling at, say, 100 m.p.h. in still air. Many researchers estimate that the aerodynamic force generated could easily reach about 250 lbf (1112 N), i.e. 20 to 25% of the car weight pressing the wheels on to the road. The aerodynamic lift acting on the rear axle of the same vehicle could be of the order of 150 lbf (667 N) [33]. The cross-wind conditions, often met on modern motorways, can aggravate the effects of aerodynamic lift by boosting it during a wind gust, thus creating dangerous response and steering problems.

In aeronautics, wing aerofoils generate aerodynamic lift in 'free flight'. As the motion of a car takes place in the proximity of the ground, the ground interference plays an important role in lift generation. The basic shape of a wing aerofoil is shown in Fig. 4.1(a). Note that the upper surface is curved more than the lower and consequently the airstream travels further over the upper surface than the lower in the same period of time. This causes the upper airstream to move locally faster than that along the lower surface. This results in a flow pattern which can be considered as composed of two different superimposed flow patterns [21,22]. One is simply the uniform

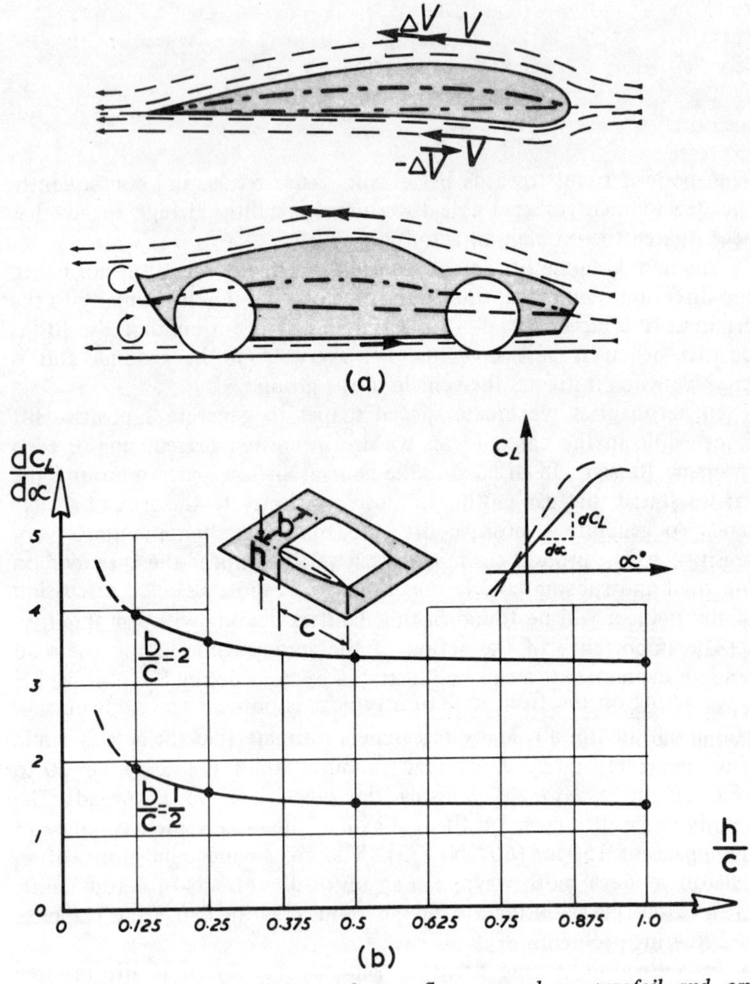

Fig. 4.1. (a) Principle of circulatory flow around an aerofoil and car. (b) Variation of the lift curve slope of a short-span wing with decreasing ground clearance

airstream past the aerofoil, with flight velocity V, while the other is a flow circulating around the aerofoil. Therefore, we have the aerofoil moving at a speed V while the air also circulates around its contour. We called this kind of motion 'circulation'. It is found that the aerodynamic lift force is directly dependent on this composite flow and proportional to the speed of 'circulation'.

The strength of this circulation is a function of the aerofoil shape and, more particularly, a function of the difference between the curvatures of the upper and lower surfaces. This curvature may be represented by the camber line (defined in Section 3.5).

The curvature of the camber line is simply called *camber* and is measured as the value of the maximum height above the chord line expressed as a percentage of the chord line length. Adapting our experience in aeronautics to the case of a road vehicle we may also expect aerodynamic lift to depend on circulation and on the curvature of the camber line of the vehicle shape, Fig. 4.1(a).

4.1 GROUND EFFECT

The problem of circulation is somewhat enhanced in the case of a car which, moving close to the ground, has the underside flow slowed down due to interference with the viscous boundary layer. This increases the circulation and further increases the upward aerodynamic lift. Thus the flow pattern around a car is more complex than that around an aeronautical aerofoil in free flight. The small width-to-length ratio of a car introduces a more complex three-dimensional flow pattern than is usually found in the case of aeronautical wings of comparatively large span. This three-dimensionality means that more air tends to flow sideways around cars. Nevertheless as aircraft also move close to ground, during take-off and landing, the problem of the ground effect on wing characteristics has attracted a considerable amount of interest among aeronautical engineers who have considered an aerofoil of the same span to chord ratio as the average vehicle [31,32,43]. It was found that the value of the lift force changes in the vicinity of the ground, and also that the slope of the lift curve changes. The characteristic ratio of the width to length of a car induces air to flow sideways at a greater rate than in the case of an aeronautical wing which has a much greater span. As Fig. 4.1 shows, the slope of the lift curve increases when the aerofoil moves closer to the ground. Investigations of the behaviour of wing aerofoils in the proximity of the ground were carried out at comparatively large ground clearances as required under the conditions of aircraft take-off and landing. For conditions in which an average car normally operates we have to investigate a whole range, including very small ground clearances. Consider, for example, a simple cylindrical rod of length such that the flow around it (normal to its axis) may be considered two-dimensional, i.e. the flow pattern around it lies in a piane normal to its axis [44]. In the absence of the ground effect the pressure distribution around such a cylinder would be horizontally symmetric and, consequently, the aerodynamic lift acting on it would be zero (Fig. 4.2(a)).

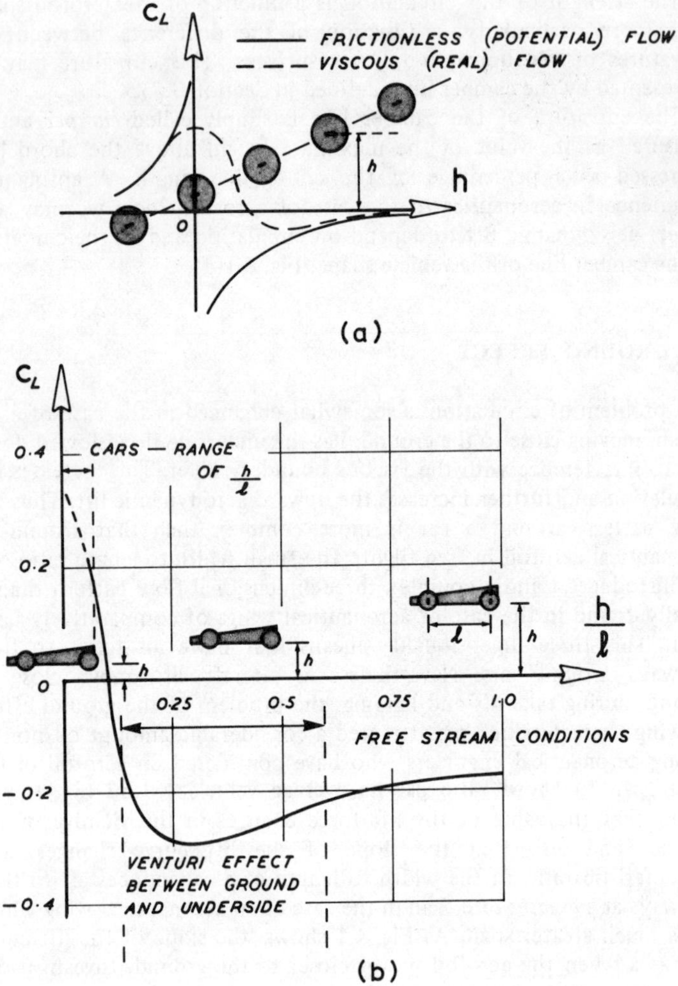

Fig. 4.2. (a) The lift coefficient variation of a cylinder gradually approaching the ground. (b) Aerodynamic lift variation of a car with gradually decreasing distance from the ground.

When we introduce the ground (represented by the *Oh* axis in Fig. 4.2(a)) and move the cylinder gradually towards it, the pressure distribution becomes increasingly asymmetric resulting in a changing aerodynamic lift force. This diagram shows the variation of the lift coefficient C_L as a function of the height h of the cylinder over the surface. The region between the cylinder and the ground presents a 'restriction' to the flow, and consequently the local flow velocity there

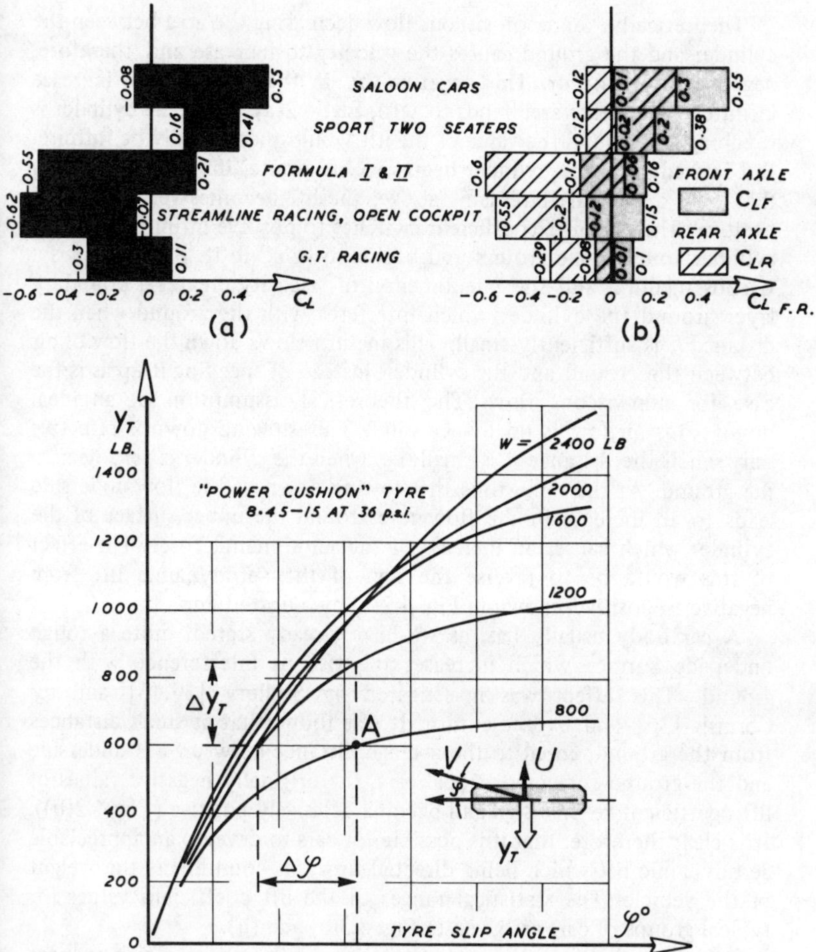

Fig 4.3. (a) Lift coefficient range for typical groups of cars; (b) front and rear axle distribution of lift coefficient range for typical groups of cars. (c) Curve of tyre side-frictional force versus side-slip angle

increases. The application of Bernoulli's law (Equation 2.1) indicates that the local pressure on the surface of the cylinder and the ground opposite decreases. The resultant aerodynamic force (expressed in the form of the coefficient C_L) would then act towards the ground and therefore its sign would be negative. In aerodynamics, this phenomenon is often referred to as the 'Venturi effect'.

Theoretically, for a non-viscous flow decreasing the area between the cylinder and the ground causes the velocity to increase and, therefore, negative lift to occur. This negative lift, in theory, tends to increase infinitely when this area tends to zero, Fig. 4.2(a). When the cylinder is touching the ground the value of the lift would theoretically be infinite. But immediately the cylinder begins to 'penetrate' into the ground, so that only a part of it remains above, the lift becomes very great and positive. Thus the lift coefficient switches to positive infinity when the cylinder touches the ground and begins to 'sink' in it. But the air is a viscous medium and the phenomenon of viscosity creates a boundary layer around the cylinder which interferes with the ground when the distance h is sufficiently small. This in turn slows down the flow of air between the ground and the cylinder, instead of speeding it up as is the case for non-viscous flow. The theoretical assumption of an ideal non-viscous airflow is no longer valid. This slowing down is effective only when the distance h is small, i.e. when the cylinder is very near to the ground. At the same time the slowing down of the flow underside leads to an increase in the flow rate around the upper surface of the cylinder which causes an increase of the aerodynamic force. The effect of this would be to reverse the sign of the aerodynamic lift from negative to positive, shown in Fig. 4.2(a) by a dotted line.

A car body usually has, as we have already stated, quite a rough underside surface which increases the viscous interference with the ground. This effect was investigated by Stollery [29,45] and by Cornish [36] and by the author. It was found that at small distances from the ground, equal to the average distance between the underside and the ground for conventional cars, the originally negative values of lift coefficient reverse sign and become decisively positive (Fig. 4.2(b)). It is clear therefore, that it is possible for cars to develop an appreciable aerodynamic lift which being directed upwards counteracts the weight of the vehicle. The statistical ranges of the lift coefficient values for typical groups of cars are given in Fig. 4.3(a) and (b).

It is the job of a good stylist to choose a car shape which reduces this lift force, or even reverses its sign. In other words, an average car at speed tends to 'float' over the road as the aerodynamic force generated by its motion counteracts the weight. It would be better if this aerodynamic force pressed the car on to the road, and thus improved the grip of the tyres.

4.2 TYRE–ROAD FRICTION VARIATION

The car stylist and designer would like to know how this aerodynamic force operates and, specifically, how it is distributed between the front and the rear axles. The car driver would like to have more detailed information about this phenomenon, since not only would it enable him to better understand his car behaviour, but also so that he would be better prepared for any kind of emergency, either at speed or in strong cross winds.

The effect of aerodynamic lift on the tyre-road frictional reaction is a typical example of how car aerodynamics is intimately mixed with the purely mechanical or dynamic problems of car motion. We cannot discuss the one without the other. The rolling of the tyre on the road can occur either in or obliquely to its plane of symmetry. In the first case we have simple rolling which, when the propelling torque is applied, creates frictional forces acting in the same plane. In the second case, the wheel not only rolls but has also a 'side-slip' velocity U, and its plane of symmetry is inclined by the 'side-slip angle' to the actual direction of motion V (Fig. 4.3(c)). This motion creates the tyre side force Y_T. This force is necessary in all steering and turning operations. The value of the side force Y_T is very sensitive to the load W which presses the wheel on to the road surface. Typical curves showing the dependence of this force Y_T on the side-slip angle, for different values of the applied load, are presented in Fig. 4.3(c).

From this diagram, considering a tyre under a typical load of 1200 lbf (5338 N) and working at a side-slip angle of say 3.5°, the actual decrease of the load to say 800 lbf (3558 N) would require an increase of the side-slip angle to about 6.4° to maintain the same side-force $Y_T = 600$ lbf (2669 N); otherwise the side-force will decrease to 460 lbf (2046 N).

The load acting on a car wheel is $(W - L_A)$ so that we have to try to reduce the value of the aerodynamic force L_A.

During the process of styling a new type of car, and when the shape has been decided, it is of considerable importance to be able to estimate not only the actual value of the aerodynamic lift force, but also its location, i.e. the theoretical point of its application (the so-called centre of pressure, C.P.). This is of particular importance because we have to estimate not only the aerodynamic force itself, but also its distribution between the front and the rear axles, and its moment about the centre of gravity of the vehicle. A knowledge of the position of the C.P. allows us to estimate the effect of the aerodynamic lift on the front axle steering characteristics and on the rear axle traction efficiency.

A preliminary estimate of the lift distribution can be obtained either from direct measurement of the lift and pitching moment on a scale model in the wind tunnel, or by integration of the measured pressure distribution on the surface of the scale model and mathematically finding its moments around an arbitrary reference point (Fig. 4.4(a)). Knowing the wheelbase distance l and the relative position of the C.P. it

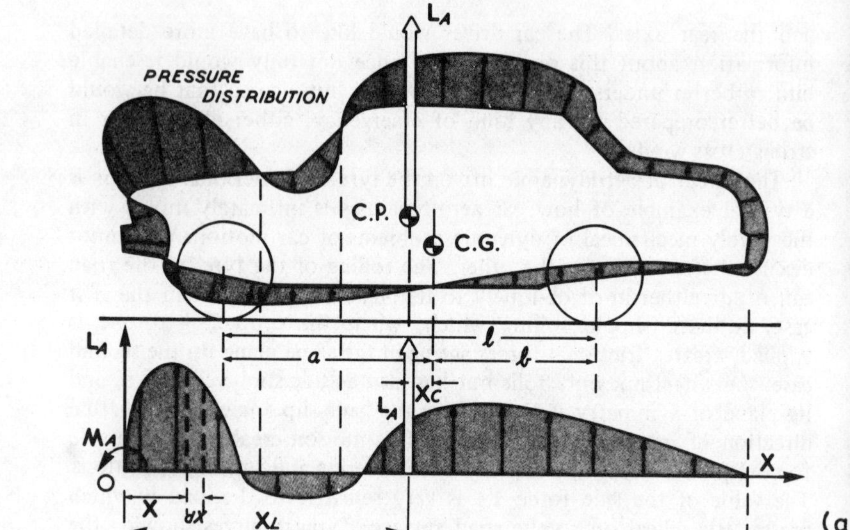

Fig. 4.4. (a) Method of estimating the aerodynamic pitching moment; (b) wheel deflection pattern and corresponding yaw angle; (c) rear view of a trailing vortex formation

is easy to calculate the actual distribution of the lift on the front and rear axles. Knowing this distribution, we can estimate the actual effect of the aerodynamic lift force on the individual wheel loads and, consequently its importance in tyre-road interaction at speed.

In aeronautical wing theory the generation of lift is related to circulation; the body shape generates the lift by producing a circulatory airflow. As was previously stated when discussing the general flow pattern around the car, we can represent a circulatory flow by a 'vortex', remembering that this vortex does not terminate, but continues as a 'trailing vortex' downstream (Fig. 4.4(c)).

The total aerodynamic lift force usually acts in front of the centre of gravity of the car. Thus the lift on the front axle is usually greater than that acting on the rear axle. This is due to the general shape of the engine bonnet and raked windscreen which facilitates a smooth airflow and produces a greater suction on the upper surface than the downward suction due to the flow underneath the vehicle. The rear parts of the car are affected to a greater degree by flow separation and the wake–the resultant vertical component of the aerodynamic force being much smaller. Thus aerodynamic lift primarily creates a steering and instability problem when the aerodynamic side-force acting in front of the centre of gravity tends to swing the front of the car from the direction of motion (Fig. 4.4(b)).

4.3 REDUCING LIFT BY STYLING

The reduction of the aerodynamic lift, from the safety aspect, is probably more important than the reduction of the aerodynamic drag which influences car performance and economy. It is therefore useful to examine the factors which determine the aerodynamic lift and the relationship between the choice of particular styling features and the resultant lift characteristics.

In doing this one can easily get involved with a multiplicity of details of styling which interact, often producing conflicting influences. There are nevertheless some fundamental principles to follow and only these will be considered. However, even basic principles of styling are often contradictory and in the Author's experience there seems to be lack of agreement in some of the published literature on this subject [3]. The reason for a lack of a clear and precise code of practice in relation to styling, with a view to reducing the aerodynamic lift, is probably due to the fact that we are dealing with the composite effects of features which are particular to a given car and this tends to cloud the issue.

Investigators have designed a composite model in which the features such as the comparative dimensions of the bonnet, the canopy, the boot, the slope of the windscreen and rear windows, the slope of the front underside panel, etc. were made variable [34]. The resultant lift coefficients are shown in Fig. 4.5(a).

It is noticed that an increase in length of the canopy (expressed here in terms of the ratio to wheelbase length) from 0.93 to 1.17 causes the total lift coefficient of the car to decrease considerably. On the other hand investigations seem to indicate that a decrease of the angle of the frontal underside panel from 10° to 0° also results in a decrease of the aerodynamic lift coefficient. This is in agreement with results obtained by other workers [87].

The relative position of the nose of the vehicle with respect to the ground influences the vertical division of the airflow. A highly elevated nose, as shown in Fig. 4.5(b), tends to reduce the mass airflow rate over the upper surface, and at the same time, directs the flow into the gap between the underbody and the ground.

A low position of the nose tends to induce the mass airflow to move up over the upper surface. As the flow between the underbody and the ground is affected by the interaction of the viscous boundary layer, and is therefore restricted, part of the airstream forced underneath the vehicle will tend to flow sideways. It will consequently increase the strength of the trailing vortices and the aerodynamic lift force.

The problem of the correct positioning of the nose cannot be solved separately. Other features of the vehicle shape interfere with the airflow pattern around the front parts and have to be considered. The angle and the curvature of the windscreen, and the overall ground clearance are the main influential features. A well-inclined and rounded windscreen helps in maintaining a smooth airflow over the greater part of the upper surface. This in turn tends to lower local pressures there and leads to an increase in lift. On the other hand, the increased ground clearance helps in maintaining some sort of continuous airstream underneath, which consequently results in lower local air pressures and in a decrease of the total aerodynamic lift.

Styling the car underside

It would be beneficial to try to increase the mean airflow velocity in the gap between the underside and the ground. Theoretically, we should aim at creating the 'venturi' type of flow which would consequently produce a downwards suction, i.e. a negative lift force.

Vehicles should have their centre of gravity as low as possible and consequently the underside panel then has only a small ground clearance. This, however, still allows longitudinal grooves to be let into the underside panel to facilitate the airflow between the ground and the car. The relief of the flow blockage underside would help to lower the pressure there and thus would tend to reduce the aerodynamic lift on the car.

Most vehicles have very rough undersides and it is of interest to consider the effect of fitting an underbody panel [46]. Statistical analysis indicates that fitting such a panel can reduce the lift coefficient (at zero angle of yaw) by approximately 20%. The side edges of the underside panel should have sufficient sidewise curvature to allow

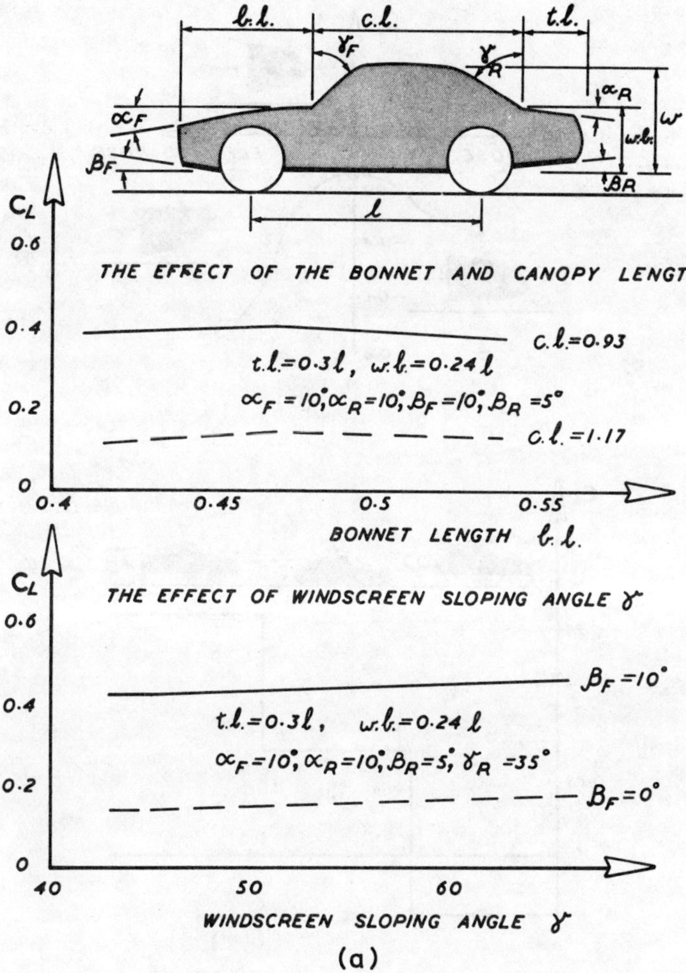

Fig. 4.5. (a) Effect of styling proportions on the aerodynamic lift coefficient

pressure equalisation between the underside and the region along the sides of the vehicle. A car in motion creates a turbulent wake which due to viscous effects tends to move with the vehicle. Inside this wake there are negative local pressures. By allowing these pressures to act in the region between the ground and the underside we lower the local pressures on the underside and thus decrease the positive aerodynamic lift of the car.

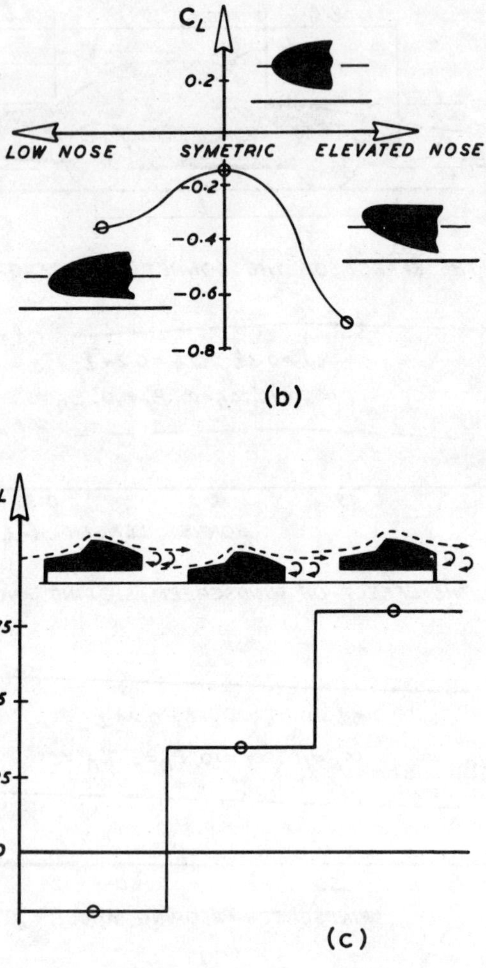

Fig. 4.5. (b) Effect of the position of the car nose on the aerodynamic lift coefficient; (c) effect of blocking the underside airflow on the aerodynamic lift coefficient

Experimenting with a car model of a typical shape and obstructing the underside by (1) sealing the ground clearance in front of it (Fig. 4.5(c), (2) setting the model directly on the ground, and (3) obstructing the underside flow at the rear, are three extreme situations. Opening the underside region to the wake helps to lower the pressure underneath and produces a negative lift which prevails over the positive lift created by the flow over the upper surface. Sealing the ground clearance from the rear connects the underside region with the comparatively high pressure region (stagnation) in front of the car. This results in an increased pressure which produces increased positive lift.

All these experiments are of rather a hypothetical character since a total blockage of the flow either in front or at the rear would not occur in practice. In reality we can only consider a partial blockage, due to such items as a specially designed under lip in front of the front axle, or the especially low positioning of the front number plate, or by fitting rubber mud flaps.

Modern cars can be divided into distinct typical shapes. The saloon car is produced either with a fast streamlined body shape, or as a 'family saloon car' with a distinct bonnet and rear boot configuration. On the other hand, we have the estate car type, distinguished by a long cabin and a flat rear surface.

All these types differ mainly in the styling of their rear parts, in which the back rake and shaping of the rear part of the canopy is the important feature. The length of the boot and the inclination of the rear window have a decided influence on the aerodynamic lift, especially on the component part which acts on the rear axle. Lengthening the rear boot, while maintaining the rear window angle at approximately 45°, results in a reduction of the rear axle lift. Increasing this angle so that the top of the canopy is smoothly joined to the rear edge of the boot forms a streamlined shape, colloquially known as a 'fast-back'. Fast-back styling tends to maintain a comparatively smooth airflow over the rear which, helped by the side trailing vortices, tends to 'push' the airflow down. This results in an increase of the upward directed aerodynamic lift which acting on the rear axle, upsets good frictional interaction of the rear wheels with the ground.

Aerodynamic effect of elevating the front and rear ends

A study of pressure distribution on the upper and underside of the car helps in an analytic approach to methods of reducing lift. The aim is to reduce the negative (suction) pressure on the upper surface and also any positive pressures on the underside. Consequently, we should avoid such styling which will obviously tend to increase lift, i.e. styling which will increase the pressure underneath or increase suction over the upper surfaces.

It is known from the theory of aeronautics that the lift generated by a given wing profile is optimised when the profile constitutes an 'infinite span' wing, i.e. when the airflow in which it is immersed is plane and two-dimensional. Any finite wing, of the same aerofoil shape, will be less effective in generating lift [15,16,36]. This rule is best illustrated (Fig. 4.6(a)) by a curve depicting the variation of the lift coefficient with the angle of incidence for the case of a short-span wing, with and without the wing-tip ends. The wing-tip ends restrict the sideways flow and thus create an airflow pattern which is approximately two-dimensional, i.e. similar to that over an 'infinite span' wing.

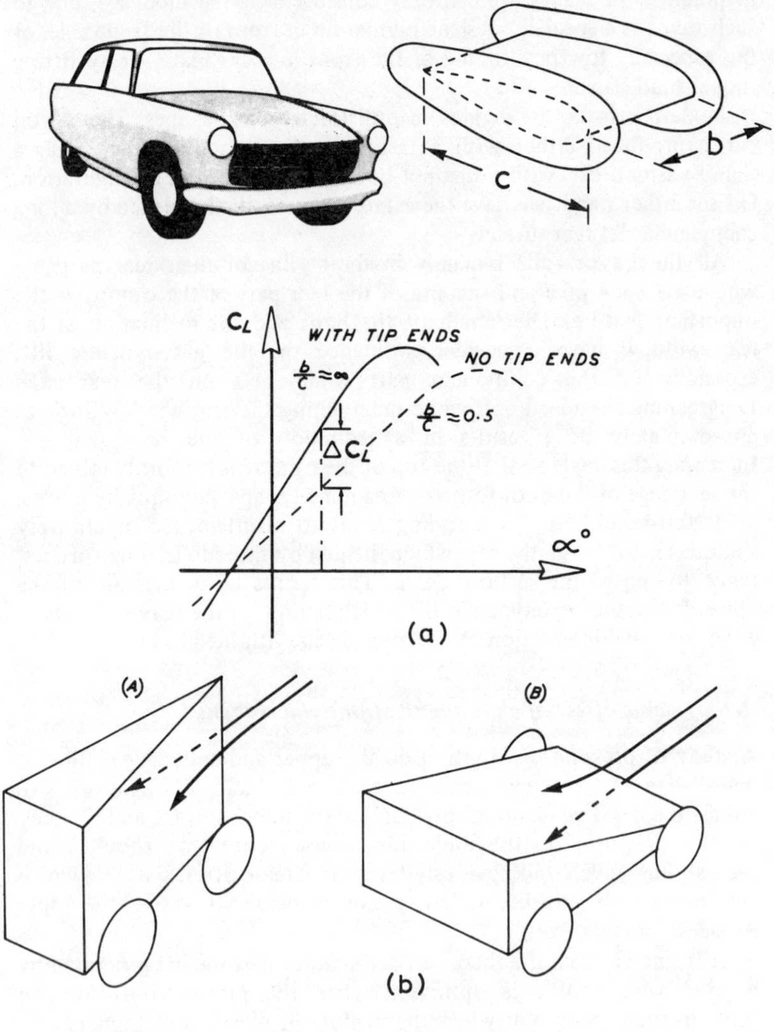

Fig. 4.6. (a) Effect of elevated headlamps and high rear side-lights on the aerodynamic lift. (b) Trends in car styling; vertical and horizontal wedge

Any styling of the vehicle shape which restricts the sideways airflow would help in augmenting the aerodynamic lift. Therefore elevated headlamps and mudguards in front, and side fins including rear lamps at the back, could be compared to the 'wing-tip' ends which effectively increase the aerodynamic lift.

The car planform

The effect of such artificial 'end plates' can be qualitatively explained as acting as some sort of air scoop which increases the mass-flow rate over the bonnet and then on to the windscreen. This increases the local airspeed with a consequent decrease of pressure (suction) which tends to increase the aerodynamic force. To avoid this, the stylist should shape the frontal part of the vehicle so that the airflow is divided, more flowing sideways than vertically, Fig. 4.6(b). Dividing it sideways (as was first done in about 1930) does not affect the horizontal asymmetry of the pressure distribution, while a vertical division adds to the generation of the aerodynamic lift.

This flow pattern affects the choice of the general planform of the car; a triangular shape adapts better to this principle than the rectangular one at present so fashionable.

Conclusions

The relative position of the nose of the vehicle with respect to the ground, and to the rest of the body shape, influences the vertical division of the airflow. Cars with an elevated nose (Fig. 4.5(b)) tend to have a reduced mass flow rate over the bonnet and an increased flow through the gap between the underbody and the ground. On the other hand a low nose position tends to induce the airflow to move up and over the bonnet. As the flow underneath is always restricted, to a greater or lesser degree due to the viscous effect, any superflous airflow forced under the car causes sideways overflow. This, as was stated earlier, increases the formation of trailing vortices and by building up the pressure underneath, increases the upward aerodynamic lift. It was found that the actual styling of the frontal profile of the car, especially the part leading to the underside, is of great importance. The qualitative tests of the effect of the nose styling is shown in Fig. 4.5(b).

The problem of the correct positioning of the height of the nose cannot be considered separately from other general features of the car shape which influence the general flow pattern and thus indirectly change the effectiveness of the nose styling. The angle and the curvature of the windscreen, the length of the bonnet and the ground clearance seem to be the main influential features. A smooth and continuous airflow over the greater part of the upper surface is helped by a windscreen which is inclined well backwards and well rounded off.

This in turn lowers the average pressure and consequently increases the aerodynamic lift.

Opening the underside to the wake and thus lowering the pressure there can be achieved by elevating the underside gradually towards the rear, for example, as in the E-type Jaguar. This leads to a lowering of the pressure on the underside surface so that the car is 'sucked' towards the ground thus reducing the upward aerodynamic lift. Smoothing the underside of the car also helps to unblock the airflow with the same beneficial result [133]. Fitting a smooth panel can reduce the lift coefficient by about 20%. The rounding off of the side edges of the underside (sills) also assists in equalising the pressure between the sides and the underside and hence in reducing the aerodynamic lift.

The importance of reducing the aerodynamic lift has already been stated. It is important in maintaining good tyre-road adherence. Thus aerodynamic lift, which is dependent on the car shape and not its weight, is more important for comparatively light cars than for heavy cars. The importance of aerodynamic lift reduction is therefore somewhat relative.

Since aerodynamic lift varies with speed it is not the lift coefficients which we have to compare but the actual aerodynamic forces. The road holding characteristics of light and fast cars are easily affected at high speed as the actual aerodynamic force is proportional to the square of the air velocity. This can happen even when the lift coefficient of the car is reasonably low, i.e. when the car has comparatively good aerodynamic characteristics. This is why, for high performance cars, designers and stylists are not satisfied simply with low lift characteristics, but go further and try to reverse the sign of the aerodynamic lift, thus increasing the actual load on the wheels. This trend is now prevalent.

4.4 SPOILERS AND NEGATIVE LIFT DEVICES

Fast cars, such as sports and racing cars, which have comparatively small positive aerodynamic lift coefficients, suffer considerably from the aerodynamic lift force due to their high speeds. Designers of these types of vehicles are now employing special lift reducing methods. Some of these methods cannot be fully adapted to saloon car design, but some are of a general nature and may be eventually applied. All the methods are concerned with the airflow over the upper surfaces of the vehicle. Devices for reducing, or even reversing, the sign of the lift can be divided into the following two categories:

(1) Spoilers, affecting the airflow pattern, and reducing the lift by modifying the flow over the upper surface.

(2) Negative lift wings, mounted above the vehicle and generating negative aerodynamic lift, i.e. directed downwards.

One can occasionally see even well known car makes adopting these devices not so much to improve the aerodynamic lift characteristics, but rather as a styling gimmick. The fact that these devices are used on racing cars has made them fashionable and attractive to fast car owners, even though as a styling feature such devices are aerodynamically useless because of the car shape. It is possible to observe a device whose obvious role is to produce an aerodynamic force, fixed behind the rear window and totally immersed in the turbulent wake behind the car, which makes it inoperative. In this way some fast car enthusiasts only display a deep ignorance of the laws of aerodynamics.

It is useful to investigate airflow conditions in which either spoilers or negative lift devices work reasonably effectively in reducing the aerodynamic lift.

Spoilers

The purpose of spoilers, by definition, is to modify the airflow around the car in such a way that the aerodynamic lift is reduced. If possible, this should be done whilst simultaneously reducing the value of the aerodynamic drag. Generalising the problem, we may say that a smooth, fast local airflow results in lower local pressures, which tends to 'suck' the car up, increasing the lift. The role of the spoiler is to modify this smooth, fast flow by slowing down the airflow over the upper surface thus increasing the pressure and consequently decreasing the lift, or by choking the flow through the gap between the underbody panels and the ground and generating lower pressures and hence downforce in this region.

There are three basic airflow regions to be considered, that over the bonnet and windscreen, that over the rear window and boot, and that under the front panel. At the front of the car the windscreen itself may act as a spoiler, depending on the angle the windscreen makes with the bonnet, Fig. 3.2(a) and (b). The separation bubble which forms over the corner between the windscreen and bonnet, when the inclination is greater than the critical angle, causes the flow to 'pile-up' at this point locally increasing the pressure. Increasing the pressure causes a reduction in the upward lift in this region. Thus the windscreen effectively acts as a 'spoiler'. Streamlined cars having windscreens rather well raked back do not obviously use them as spoilers.

Some time ago racing cars used a spoiling device called 'the Ferrari ridge', having been first used by the famous Italian car maker, Fig. 4.7(a). This type of device is fixed well back, usually at the rear edge of the car in the region where the airflow is about to separate. This is near and upstream of the wake. By arranging a ridge in this region we induce the slowed down airstream to separate in front of it (similar to

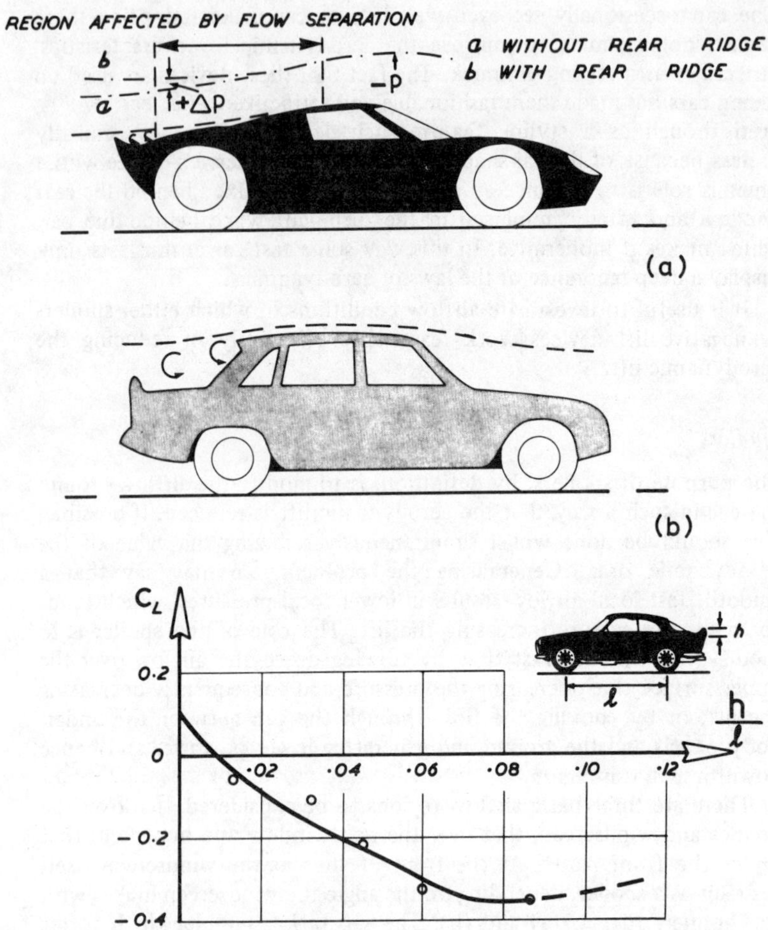

Fig. 4.7. (a) Sports car with 'Ferrari ridge'; (b) rear ridge overshadowed by canopy; (c) effect of rear ridge height on the aerodynamic lift

the case of the bonnet and windscreen configuration discussed in Chapter 3). This causes the local pressure in the region in front of the ridge to increase due to the formation of a large 'separation bubble'. The pressure over a large area at the back of the car is increased, thus reducing the overall aerodynamic lift there. The airflow, triggered to separate from the upper surface of the 'fast-back', is re-attached to the upper lip of the ridge. This brings an additional benefit by having more stable aerodynamic conditions over this part of the car under varying conditions of speed and relative wind. The effectiveness of such a ridge depends on its height, as this affects the extent of the separation area. The graph in Fig. 4.7(c) shows the dependence of the lift coefficient C_L on the ridge height.

Preliminary experiments on models in a wind tunnel indicate that there is an optimal height for the ridge, but its effectiveness depends also on the rest of the car configuration, and the 'history' of the flow before it reaches the ridge. For example, Fig. 4.7(b), the ordinary saloon car (not the fastback), has its rear window set in a frame which usually protrudes, and the flow separation may be triggered at its upper edge. A 'Ferrari ridge' placed at the rear edge of the boot would be then fully immersed in the turbulent wake making it quite useless as a spoiler. No re-attachment of the flow can occur and its presence would have no effect in changing the local pressure. It seems therefore that many cars with some type of a 'ridge' at the back are in this category–it then becomes a styling (or sales) gimmick and ceases to be a real spoiler.

The effect of a rear spoiler on a sports car is somewhat similar to that which a flap has on the underside or pressure surface of an aerofoil in that it increases the pressure on the surface immediately ahead of it. Depending on the inclination of the rear body panels, this increase of local pressure coefficient, which may be by as much as 0.2, can reduce both the drag and the lift at the rear axle by about 0.05 and 0.15 respectively for a ridge between about 1.5 in (40 mm) and 3 in (75 mm) high. This type of spoiler has no detectable effect on the flow and pressure distribution acting over the forward portion of the vehicle, and so it changes neither the contribution to total drag which arises from the front end, nor the lift acting at the front axle. However, the ridge spoiler does affect the distribution of the net lift force between the front and rear axles.

The front spoiler or airdam is of increasing popularity with designers of saloon and sports cars of all sizes. The detailed design of the best shape for a particular vehicle may involve some trial and error work in the wind tunnel, and is generally optimised with the cooling air flow blocked off. The front spoiler is often mounted either under the front panel or may form an integral moulding with the front bumper. Its principles of operation are quite well understood. This spoiler reduces the airflow through the underbody region by choking or damming the airflow at the front end. By reducing the airspeed in the region in the lee of the spoiler, in the area which is adjacent to the rough underside surface and in the region around the front suspension, the drag of the underbody region is reduced. If the front spoiler is properly matched to the car and the detailed design of the front end is treated as an entity, this drag reduction can exceed the additional drag which originates from the pressure difference acting across the spoiler. A front spoiler causes a reduction in pressure on the front underbody which, in addition to improving the flow of cooling air and reducing the net drag of the vehicle by up to about 0.01, can reduce the lift over the front axle by up to 0.05. This type of spoiler has very little or no effect on the flow over the rear of a full-scale car on the road and so does not change the lift acting at the rear axle. Care must be taken in scale model tests in wind tunnels as the reduction in the ground clearance can induce erroneous effects if the growth of the boundary layer on the floor of the wind tunnel is not correctly accounted for.

Negative lift devices

Increasing demands of the racing track, especially of large racing cars, now reaching speeds of the order of 320 km/h, have turned the mind of designers toward improving, if possible, the adherence of car tyres to the road. The usual spoilers were inadequate–something more drastic than simply reducing aerodynamic lift was required. Thus the idea, first advanced by American Chapparal, of reversing the sign of the aerodynamic lift by using a wing shaped device fixed over the car was born. Fig. 4.8 shows an early design of such a wing fitted to a Lotus Formula I racing car.

It was found that the negative aerodynamic lift greatly improved tyre cornering characteristics and thus permitted cars to turn at higher speeds. Similarly, the improved tyre-road interaction of the driving wheels allowed much more efficient use of the engine power when accelerating and decelerating. These problems are acute on the racing track but they are important also at all speeds and for all types of car.

The negative lift device was composed of an 'aeronautical' wing fixed to struts over the car at a negative angle of incidence. These wings fixed over the rear axle seemed to be a radical and effective solution. A car with a comparatively small lift coefficient, at about 160 km/h, can develop an aerodynamic lift force which constitutes a high proportion of its weight, while with a typical negative lift wing it can generate a downward aerodynamic force considerably greater than the original lift produced by the car shape. Thus, in effect, we can have a resultant 'negative lift' which is additive to the weight of the car.

These devices have their snags. A wing has to be fixed relatively high over the car so that it is in a turbulence-free airstream. In this position the drag due to the wing (which is of the order of 10% of the lift) acts on a long 'lever arm' h_{RW} which tends to rotate the vehicle about the point of contact of the rear wheels with the ground. This moment tends to lift the front wheels off the ground, adding to any aerodynamic lift

Fig. 4.8. Racing car equipped with a high-mounted rear negative-lift wing (By courtesy of Team Lotus Ltd)

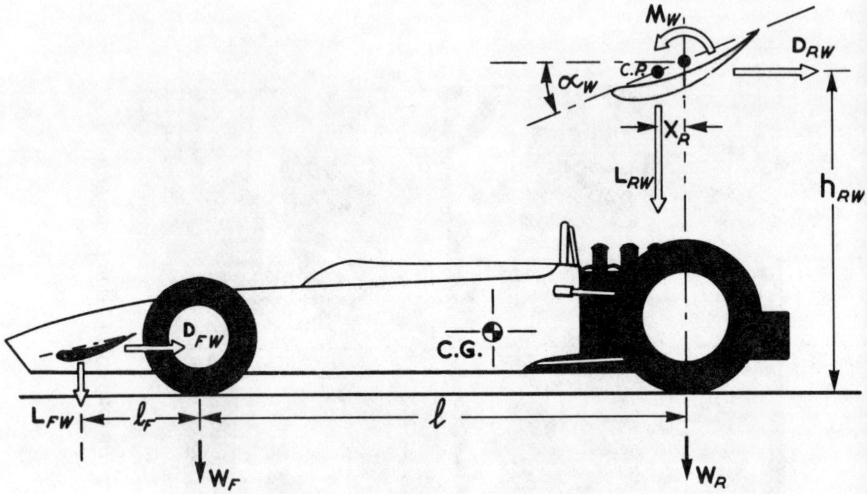

Fig. 4.9. Relative position of front and rear negative-lift wings

force already affecting it. This is important as the front wheels are steering.

Some racing car designers have got round this problem by adding another negative lift wing (Fig. 4.9) over the front axle. The result is that the total additional aerodynamic drag of both these wings greatly affects the overall performance, particularly the acceleration on a straight run. The problem of the effect of aerodynamic drag on car performance is discussed in detail in Chapter 5.

Ideally, racing car designers should seek a way to attach the negative lift device directly on to the wheel axles, thus, by-passing the suspension system. This is theoretically correct but very difficult to design in practice. Therefore most cars have these wings fixed to the main body of the car. This is inconvenient because the suspension system then acts differently every time the car changes speed, or the atmospheric wind causes a change in the relative airstream direction and velocity. At present commonly used negative lift wings are fixed near the front axle, usually a little in front of it, and the second wing is fixed over, or slightly to the front or back, of the rear axle, Fig. 4.9. The positioning of these wings depends on whether their effect is to be directed to the traction or the steering of the car.

Positioning the negative lift wing

The positioning and consequent airflow conditions of the front stub-wings and the rear wings are basically different and need to be separately discussed.

Let us first consider the rear wing device. It serves to increase the aerodynamic downward force which is applied to the tractive rear wheels. It is convenient therefore to place it above the rear axle. At larger negative angles of incidence, nevertheless, the drag component creates a considerable nose-up moment. This moment must be compensated, at least partially. This can be done by placing the rear wing slightly upstream of the rear axle so that the resultant aerodynamic lift gives a compensating moment.

The actual effect of the wing positioning in relation to the rear axle was investigated experimentally by the Author on a typical Formula I racing car model [12]. It was found that moving the wing upstream of the rear axle tends to increase the lift slightly. This seems to be consistent with the flow pattern around the car. The local airflow is directed in a more upward way over the front part of the vehicle which causes a decrease of the effective wing angle of incidence and consequently a decrease of aerodynamic lift.

The lowering of the wing towards the car produces increasing interference between the wing and the car body. The airflow between them, although accelerating and thus creating a low pressure region on the lower surface of the wing, also creates a suction on the upper surface of the car body, the one practically cancels the other. Experimental investigation has shown that the wing effectively loses its aerodynamic efficiency below a certain height from the car body. This position is described by the ratio of the height, h, of the wing over the upper surface of the car to the car wheel base, l

$$\frac{h}{l_{crit}} \leqslant 0.62$$

For a vehicle having a wheelbase of 3 m the critical height of the wing is of the order of 1.8 m above the ground, or approximately 76 cm above the upper surface of the car [12].

The devices, originally used on racing cars and fixed at first on comparatively high struts, were quite effective from the aerodynamic point of view as they were moving in a free, undisturbed airstream above the car. But, in practice, the struts supporting these wings proved to be unstable and were effected by vibration. There were a few accidents. This prompted the International Automobile Federation to regulate the design of these devices. It was decreed that wings should be so placed that the highest point of the wing should be not higher than 80 cm above the sprung underside of the car. At the same time it was required to make these wings a structural part of the main body of the car. The car designer is now compelled to fix the rear wing very near

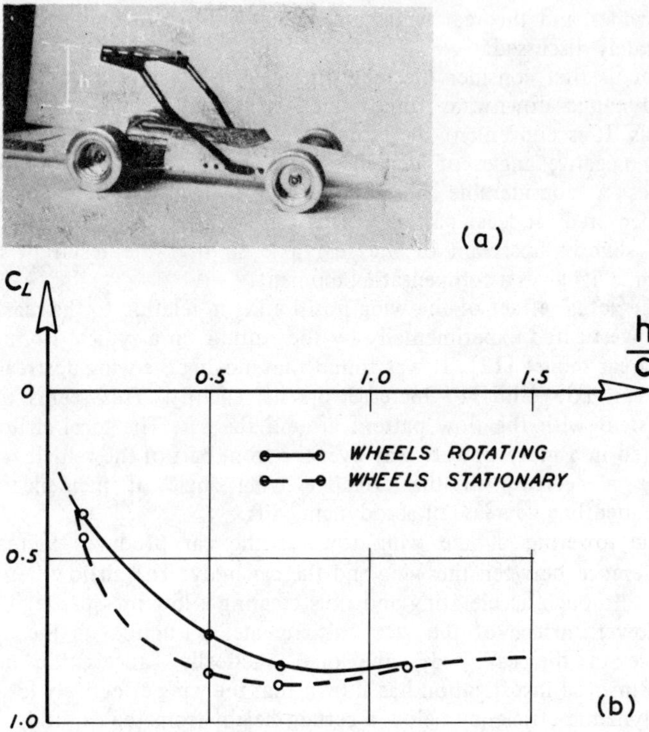

Fig. 4.10. (a) Model of a racing car equipped with an experimental rear negative-lift wing; (b) effect of wheel position and rotation on the aerodynamic effectiveness of negative-lift wing

the upper part of the vehicle. It is therefore necessary to investigate the aerodynamic effects of such a position. Experimental investigations carried out in the Handley Page Laboratory of the City University by the Author [12] have shown that the gradual lowering of the wing resulted in rapid deterioration of the negative wing lift. The graph in Fig. 4.10(b) shows the value of the wing lift coefficient as a function of the ratio *h*/*c*. Decreasing the height causes the negative lift coefficient to decrease, but when the wing becomes quite near to the car it rapidly increases. The experiment described was carried out on a model with rotating wheels. The wheels were made to rotate with a circumferential speed equal to the air speed in the tunnel. This simulated the actual car motion in still air. As the diagram shows the wheel rotation makes a considerable difference, reducing the resultant value of the lift coefficient. This is best explained in terms of the circulation principle. Air is induced to circulate with the wheel due to viscosity and this

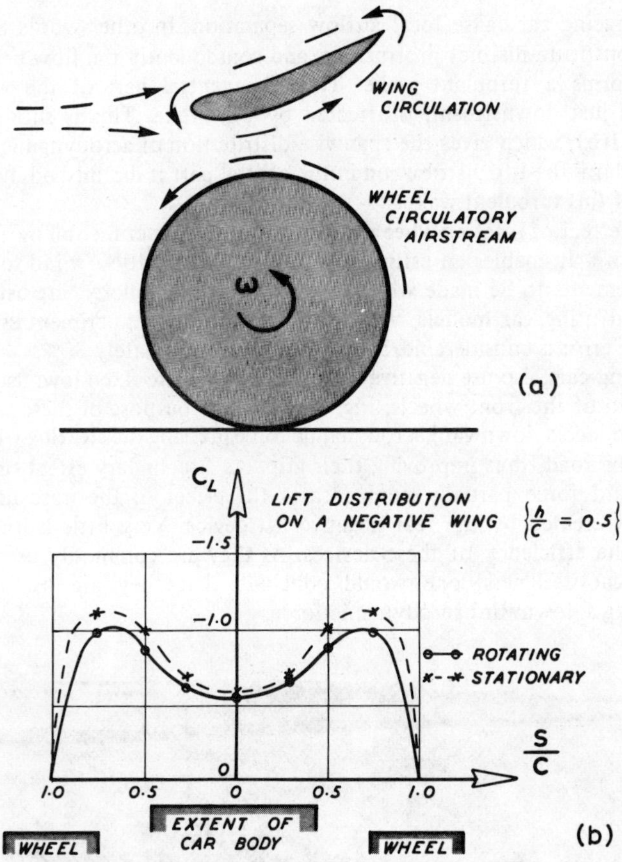

Fig. 4.11. (a) Schematic pattern of the circulatory airflow around a negative-lift wing and wheel. (b) Spanwise lift distribution on a negative-lift wing

interferes with the circulation around the wing, thus reducing its aerodynamic lift, Fig. 4.11(a).

The limitation of the wing position above the car, imposed by the International Racing Authority, places it at approximately

$$\frac{h}{c} = 0.45,$$

which is comparatively very low. This results in the device being less effective in generating negative lift. One of the reasons for this is the interference of the main car body in front of the centre part of the wing. This includes the driver's compartment and windscreen, as well as the engine which is usually situated just behind. These structural parts of the racing car cause local airflow separation. In other words these parts constitute distinct protrusions, and consequently the flow behind them forms a turbulent wake. Thus the central part of the wing, situated just downstream, is affected by this wake. This is shown in Fig. 4.11(b), which gives the spanwise distribution of aerodynamic lift. We see how the lift distribution in the central part is diminished by the effect of this turbulent wake.

The effect of rotating wheels measured on the experimental model is illustrative. It enables an estimate of the actual validity of wind tunnel measurements to be made which, for reasons of simplicity, are usually obtained using car models with stationary wheels. Experiment shows that the error is considerable, amounting to approximately 30%.

Racing cars also use negative lift devices which are fixed low, usually upstream of the front wheels, Fig. 4.12(a). The purpose of these wings is to produce a downward aerodynamic force pressing the steering wheels on to the road, thus improving their grip. As a secondary effect such a downward force partially compensates the effect of the nose lifting moment created by the rear negative lift device. Very little is known about the efficiency of these devices. As they are commonly used by racing car designers, one would conclude that they are useful in providing a downward aerodynamic force.

(a)

Fig. 4.12. (a) Racing car with negative-lift wing integrated with the body shape (By courtesy of Team Lotus Ltd). (b) Spanwise lift distribution on the front negative-lift wing; (c) negative lift coefficient of the front wing as a function of the angle of wing incidence and ground clearance; (d) flow pattern due to the interference between the front wing and front rotating wheels

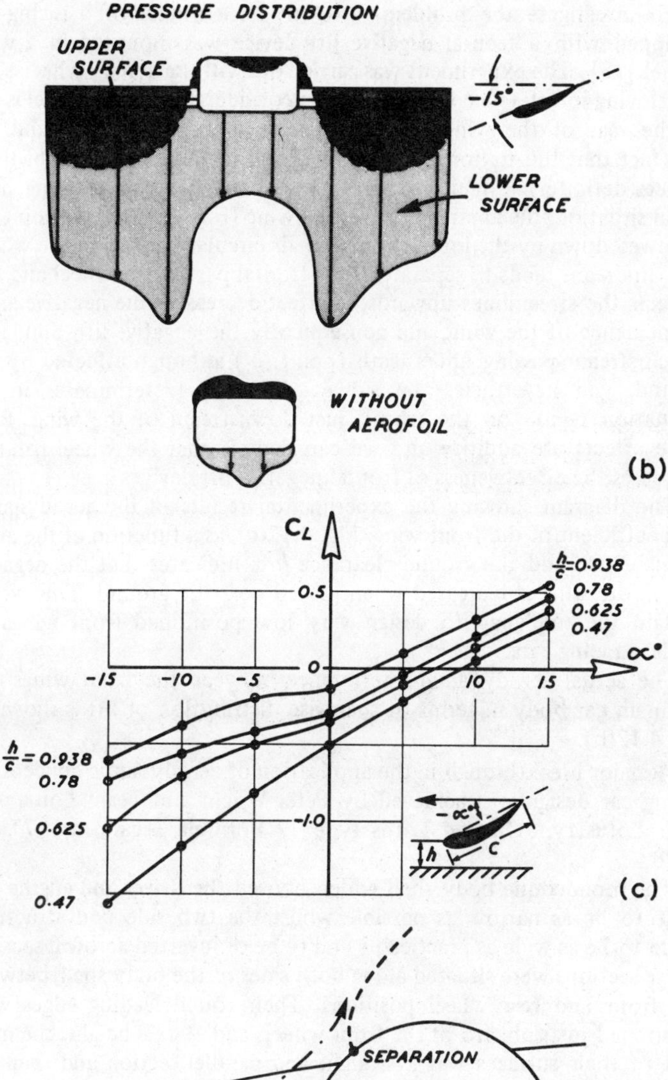
PRESSURE DISTRIBUTION
UPPER SURFACE
-15°
LOWER SURFACE
WITHOUT AEROFOIL
(b)
C_L
h/c = 0.938
0.78
0.625
0.47
0.5
0
-1.0
α°
-15
-10
-5
10
15
-α°
c
h
(c)
SEPARATION
STAGNATION
a
b
ω
(d)

To investigate the problem, a 1/8th scale model of a racing car equipped with a frontal negative lift device was mounted in a wind tunnel [13]. The experiment was carried out with stationary wheels. The wheel-wing interference which could be considerable, as the wheel is just to the rear of the wing, was not investigated, Fig. 4.12(b) and (c). The fact that the stationary wheels were situated downstream of these devices definitely influenced the results. When wheels are rotating, as in a real situation, the airstream above the wing (Fig. 4.12(d)), at point '*a*', is slowed down by the interaction with air circulating around the wheel. The airstream tends to separate in the frontal part of the wheel and this deflects the streamlines upwards, in effect decreasing the negative angle of incidence of the wing, and consequently the negative lift. Similarly, the airstream passing underneath (point '*b*') although affected by the ground, will nevertheless be slowed down as it terminates in the stagnation region on the wheel, just downstream of the wing. Both these effects are additive and we can deduce that the wheel rotation decreases the effectiveness of frontal negative lift devices.

The diagram showing the experimental results of the aerodynamic lift coefficient of the front wing, Fig. 4.12(c), as a function of the angle of incidence and the ground clearance h/c, indicates that the negative lift of the wing is increased when placed near the ground. This would explain the tendency to design very low positioned front wings on modern racing cars.

The actual aerodynamic interference between the front wings and the main car body in terms of spanwise distribution of lift is shown in Fig. 4.12(b).

A major breakthrough in the application of aerodynamic concepts to racing car design was achieved by Peter Wright and Team Lotus with their Lotus type 78 and Lotus type 79 Formula 1 cars in 1977 and 1978.

The monocoque body shell which housed the driver and engine was built to be as narrow as possible whilst the two side pod structures made to be as wide as practicable and to be of inverted aerofoil section. These sections were situated along both sides of the body shell, between the front and rear wheel positions. Their round leading edges were positioned just inboard of the front wheels and at axle height, the main part of their surfaces was a straight and parallel section and then the upper and lower surfaces were curved upwards and away from the ground, to meet at sharp trailing edges as the rear wheels were approached. These negative lift devices were inverted wings with a thickness to chord ratio of about 12% and they effectively turned the whole of the centre section of the vehicle into a wing-car, as is outlined in Fig. 4.13(a). Their aerodynamic efficiency was greatly increased by the fitting of vertical end plates or sideskirts at their outer edges, which

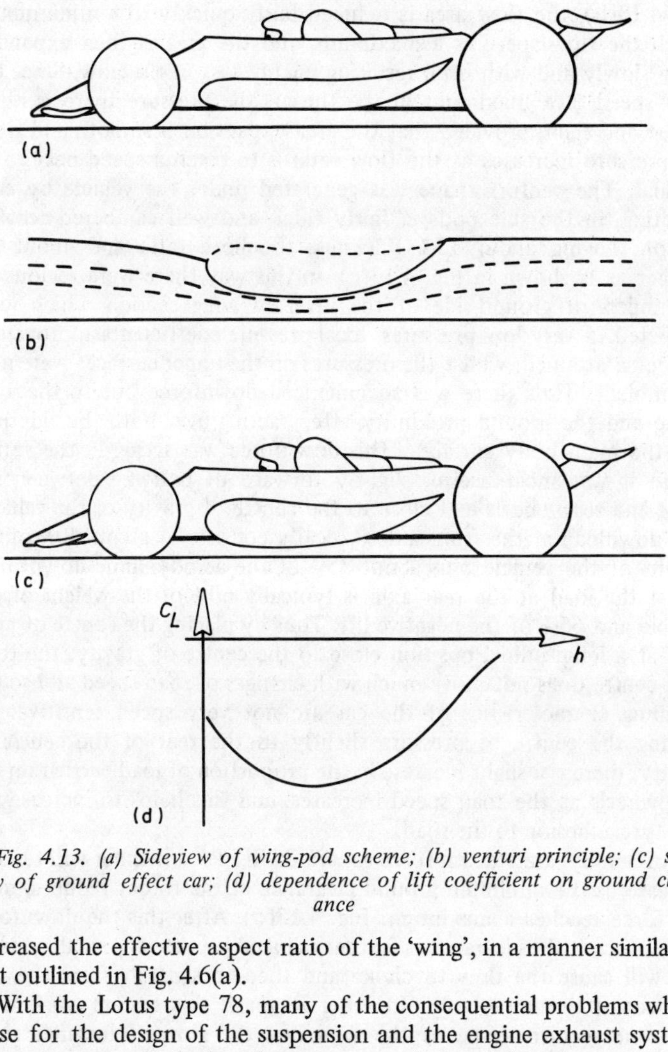

Fig. 4.13. (a) Sideview of wing-pod scheme; (b) venturi principle; (c) side-view of ground effect car; (d) dependence of lift coefficient on ground clearance

increased the effective aspect ratio of the 'wing', in a manner similar to that outlined in Fig. 4.6(a).

With the Lotus type 78, many of the consequential problems which arose for the design of the suspension and the engine exhaust system, whose components had to be kept out of the airflow over and near these 'wings', were solved and the concept was successfully proven. The major part of the car was acting as an inverted wing which was generating downwards force or negative lift due to its cambered shape and also due to the close proximity of the ground.

The Lotus type 79 was to take the ground proximity effect further and utilise the concept of the efficient shape of the venturi to generate even lower pressures beneath the body panels. In a venturi, as shown in

Fig. 4.13(b), the flow area is reduced fairly quickly to a minimum at which the flow speed is a maximum, and the area is then expanded fairly slowly and with little resulting energy loss in the flow. Since the flow speed is a maximum in the throat, the pressure there is at its lowest and then, provided that the area expansion is smooth and slow, the pressure increases as the flow retards to reach a speed near to its original. The venturi shape was generated under the vehicle by constructing, in the side pods, a fairly thick and well cambered aerofoil section, having about 13% thickness to chord ratio and about 8% camber as is shown in Fig. 4.13(c). In this way there were regions on the under- or ground-side of the inverted wing section which were subjected to very low pressures, local pressure coefficients approaching −2 being attained, whilst the pressures on the upper surfaces were near to ambient. Thus there was an enhanced downforce due to the wing shape and the ground proximity effect acting over both the side pod and the main body surfaces. This downforce was acting in the region where it was most useful, slightly forward of halfway between the front and rear wheels and close to the centre of gravity of the vehicle. The download at the front axle typically consists of about 40% of the weight of the vehicle plus about 35% of the aerodynamic downforce, whilst the load at the rear axle is typically 60% of the weight of the vehicle and 65% of the negative lift. Thus by placing the centre of pressure at a longitudinal position close to the centre of gravity, the total load centre does not move much with changes of road speed and so the handling characteristics of the car are not very speed sensitive. By placing the centre of pressure slightly to the rear of the centre of gravity, there is a slight increase in the proportion of load carried on the rear wheels as the road speed increases, and this helps to increase the rear tyre adhesion to the road.

An aerodynamic down force generated in such a manner will tend to increase as the minimum ground clearance at the throat reduces, until the force reaches a maximum, Fig. 4.13(d). After this the downforce will reduce if the clearance is further reduced because the small residual gap will cause the flow to choke and then to separate from the rear surfaces of the underbody. In order to make this system work at its best, the section shape of the wing and the ground clearance gap– typically between 2 in (50 mm) and 4 in (100 mm)– needed to be optimised, and great care had to be taken to ensure that the airflow was as clean and smooth as possible. In particular the flow over the front wing, the nose of the bodyshell and the front suspension units had to be given great attention and, most important, the flow coming over or around the front wheels had to be kept as far away as possible from the pod wing, which meant giving the front wheels a wide track. Additionally it was essential that the gap between the road and the underside of the body was sealed at the outer edges of the pods and skirts which slid up and down were used to effectively seal this gap and stop air from the 'outer' regions from being drawn in to fill the lower pressure region under the body.

The net effect of this was that at a speed of 150 m.p.h. (240 km/hr) the Lotus type 79 would generate downloads of 100 lb (445 N) on the front wings, 200 lb (890 N) on the rear wing just aft of the driving wheels, and 400 lb (1780 N) due to ground effect on the pod and body undersurfaces on a vehicle weighing 1650 lb (7500 N). Data [99] on more recent designs shows that after five years development work, the wings were together developing 600 lb (2700 N) and the underside of the car was developing 2400 lb (10,700 N) at the same speed but with a car weighing 1450 lb (6600 N).

When introduced in 1977, the Lotus type 78 had a lift coefficient of about −1.3, based on gross frontal area, and by the end of 1980 the negative lift coefficient on racing cars had been increased up to about −2.6 by a steady progression of detailed design improvements. In 1981 the regulations which had permitted the use of sliding skirts were modified by F.I.S.A., with the result that lift coefficients initially dropped back to about −1.6 but soon recovered to −2.5 inside 18 months. During this period of negative lift development, the drag coefficient of this type of car remained fairly constant at around 0.6 to 0.8, largely coming from the wheels which the regulations required to be open and uncovered, and lift to drag ratios in excess of −3 are currently being attained [99].

Whilst a reasonably low drag coefficient is useful in ensuring that racing cars have a high top speed and do not give too much power to aerodynamic drag whilst accelerating at lower speeds, as is discussed in Chapter 5, an absolute minimum drag coefficient is not an essential design objective. Top speed is often a function of the particular track on which a race is being staged and many of the major circuits have been made twistier in order to reduce top speeds and make racing safer. More important is the ability to improve acceleration, braking and cornering which results from the improved tyre-road grip which can be achieved by adding aerodynamic load to the weight load on the tyres. As an example we will consider a racing car having a mass of 750 kg and turning steadily through a corner of 660 ft (200 m) radius on a level and unbanked track. If the effective tyre adhesion factor is 1.4 then without negative lift the maximum safe road speed would be 119 m.p.h. (190 km/hr) and the car would experience a sideways acceleration of 1.4 *g*. Now add negative lift wings and a ground effect system to the car such that the product of lift coefficient and reference area is $-12\ \text{ft}^2$ ($-1.12\ \text{m}^2$), typical of the Lotus type 79. Then the maximum road speed through the corner would increase to 138 m.p.h.

(222 km/hr) and the sideways acceleration on the car would increase to 1.94 *g*. The aerodynamic download would be nearly 590 lb (2620 N) which is about 36% of the vehicle weight. Finally consider negative-lift wings and a ground effect system giving a lift-area of -24 ft^2 typical of later Lotus cars. The maximum steady road speed would increase to 169 m.p.h. (270 km/hr) whilst the sideways acceleration would be 2.88 *g* and the aerodynamic download would be 1760 lb (7810 N), or abut 106% of the weight of the car.

A number of Formula 1 racing cars rely on the ground effect downforce produced by the underside of the central section of the body to generate the basic stabilising force, and then use the fore and rear wings to set up and trim the car and its handling characteristics to suit the particular circuit and driver.

The magnitude of the ground effect force increases with the square of the speed of the car in still air, and so is most effective at high speeds but has little effect at the lower speeds which would be used in tighter corners. The venturi ground effect relied upon having a clearance gap, typically between 2 in (50 cm) and 4 in (100 mm), between the ground and the undersurface of the wing-pod, and at one stage a minimum gap of 70 mm was fixed by F.I.S.A. regulations. In order to reduce changes in the ground effect downforce which arose from changes in pitch attitude (incidence) and ground clearance height (heave), the suspensions of most Formula 1 cars were progressively stiffened until springs with rates approaching 2000 lb/in (350 N/mm) were used. Such stiff suspensions, coupled with high road speeds, gave the drivers very high vibration levels to contend with in addition to the high 'g' forces encountered in cornering, and these high accelerations interact with the detailed design of the structure [99].

It was essential that the sliding skirts sealed the gap between the wing and the ground at the outer wing-tip edge and on a number of occasions these skirts were damaged during a race which resulted in the effectiveness of the ground effect system being seriously impaired. In an attempt to reduce the dependence of ground effect downforce on vehicle speed, one constructor fitted a large fan, running in a nearly-horizontal plane, into the body of a racing car. The fan was driven from the engine and hence the reduced pressures generated beneath the bodywork were more nearly constant through the full range of road speed. This gave a major benefit to the tyre-road grip for both straight acceleration and for cornering as well as for braking. Unfortunately the fan tended to lift all loose gravel and debris from the surface of the track and hurl it rearwards, to the consternation of other drivers, and this system was rapidly ruled to be in contravention of the Regulations.

It is clear that there has been active interaction between the development of aerodynamic devices and their application on Formula 1 racing cars by the constructors and engineers on the one hand, whilst on the other hand are those responsible for the framing of regulations to govern and control motor racing as an international sport, with cars required to conform with various formulae, and with the safety of competitors and spectators in mind. The particular inter-relation between aerodynamics, engine development and regulation formation for Formula 1 was recognised by F.I.S.A. when they commissioned a computer-based study [100] of the effects that proposed changes in regulations were likely to have on racing car performance, in lap times, maximum speed attained on straight track and through particular bends, fuel consumption, braking distance and other parameters. Various changes in and constraints on the lift-area, drag-area, vehicle mass, engine and tyre characteristics were considered and a total of 24 different combinations were evaluated by computer. As a result of this study, a summary of whose results has been published [101], a new set of Regulations have been introduced for Formula 1 racing cars. Their main effect is the introduction of a full flat bottom on a car between the front and rear wheels, the banning of skirts and any systems used to fill the gap between the bodywork and the ground, and some constraint on the size and position of the rear wing. Thus it appears that F.I.S.A. have some idea of the likely effect that the introduction of these changed regulations will have initially on the performance of Formula 1 racing cars.

It seems most likely that the ingenuity of the engineers under the fierce pressure of international competition will continue to devise new ways of using aerodynamics and other sciences in improving the performance, road holding and safety of both racing and general production cars.

Chapter 5

EFFECT OF AERODYNAMIC FORCE ON PERFORMANCE

One often hears the opinion expressed that aerodynamic forces cannot significantly affect performance. It is therefore important to estimate the importance and the role of these forces in the motion of a road vehicle. To simplify the issue let us assume that a car travels at a constant speed along a straight, level road and that there is no atmospheric wind. Thus all the forces acting on the vehicle are in equilibrium and act in its vertical plane of symmetry, i.e. there are no side forces.

Apart from the aerodynamic force which can be resolved into lift and drag components, we have to account for 'tyre rolling drag' and tyre friction. The torque applied to the driving wheels by the engine produces a forward propelling force which balances both the aerodynamic and mechanical drag.

Aerodynamic force affects this state of motion in two ways. Aerodynamic lift when directed upwards counteracts the weight of the car and thus reduces the frictional adherence of tyres; the aerodynamic drag adds to the mechanical drag forces, requiring greater engine power to attain a given speed.

The frictional forces between the tyres and the road are not within the scope of this book but, for the sake of a more complete picture, they will be explained briefly. The rolling drag is caused by the asymmetry of the vertical pressure distribution on the contact surface of the tyre when rolling. This distribution gives a resultant force which is applied slightly in front of the wheel rotational axis. This results in a torque opposing the rotation, i.e. the 'rolling drag'. For further information the reader should refer to Newcomb and Spurr [47]. The tyre rolling drag is expressed by the formula

$$P_{F,R} = (W_{F,R} - L_{F,R}) f_{F,R} \tag{5.1}$$

where $W_{F,R}$ is that part of the total car weight acting on the front and rear axles, $L_{F,R}$ is that part of the aerodynamic lift acting on each axle and $f_{F,R}$ is the rolling drag coefficient of the front and rear tyres. These coefficients are dependent on the state of the road surface, on the tyre design, size, temperature and pressure. As the front wheels are often different from those at the rear we have to distinguish between the front and the rear axles.

The total rolling resistance can therefore be expressed as follows:

$$P = (W_F - L_F)f_F + (W_R - L_R)f_R \tag{5.2}$$

where the total vehicle weight

$$W = W_F + W_R$$

Adding to this the aerodynamic drag force, the total resistance, R, of a moving car can be expressed as

$$R = \frac{\rho V^2}{2} SC_D + (W_F - L_F)f_F + (W_R - L_R)f_R \tag{5.3}$$

We assume that the car moves at a constant speed and that the forces are in equilibrium. The value of the resistive force R is therefore equal to the forward tractive force T supplied by the engine through the transmission to the pair of driving wheels.

Disregarding the problem of transmission and the gearbox, T is directly proportional to the power supplied by the engine at a given speed. The power supplied is given by

$$\eta(\text{h.p.})_E = \frac{TV}{745.7} \tag{5.4}$$

where η is the efficiency coefficient dependent on transmission losses, T is in Newtons and V is in m/s. Introducing the value of T from Equation 5.4, we obtain an expression for the 'power supplied' and 'power required', consistent with the assumption of steady motion. Thus

$$\eta(\text{h.p.})_E = (\text{h.p.})_A + (\text{h.p.})_T$$

where the power related to the aerodynamic force

$$(\text{h.p.})_A = \frac{\rho V^3 S}{1491.4}\{C_D - (C_{LF}f_F + C_{LR}f_R)\} \tag{5.5}$$

and the power required to overcome the tyre rolling drag

$$(\text{h.p.})_T = \frac{V}{745.7}[W_F f_F + W_R f_R] \tag{5.6}$$

Comparing these two equations we see that from Equation 5.5, aerodynamic power is proportional to the cube of the speed, while from Equation 5.6, mechanical power is directly proportional to the speed. Plotting these two curves, and then summing them we obtain the diagram in Fig. 5.1(a). This shows that at a fairly low speed, typically about 40 m.p.h. (or 65 km/h) for an average saloon car, the power required to overcome mechanical drag is about the same as the power

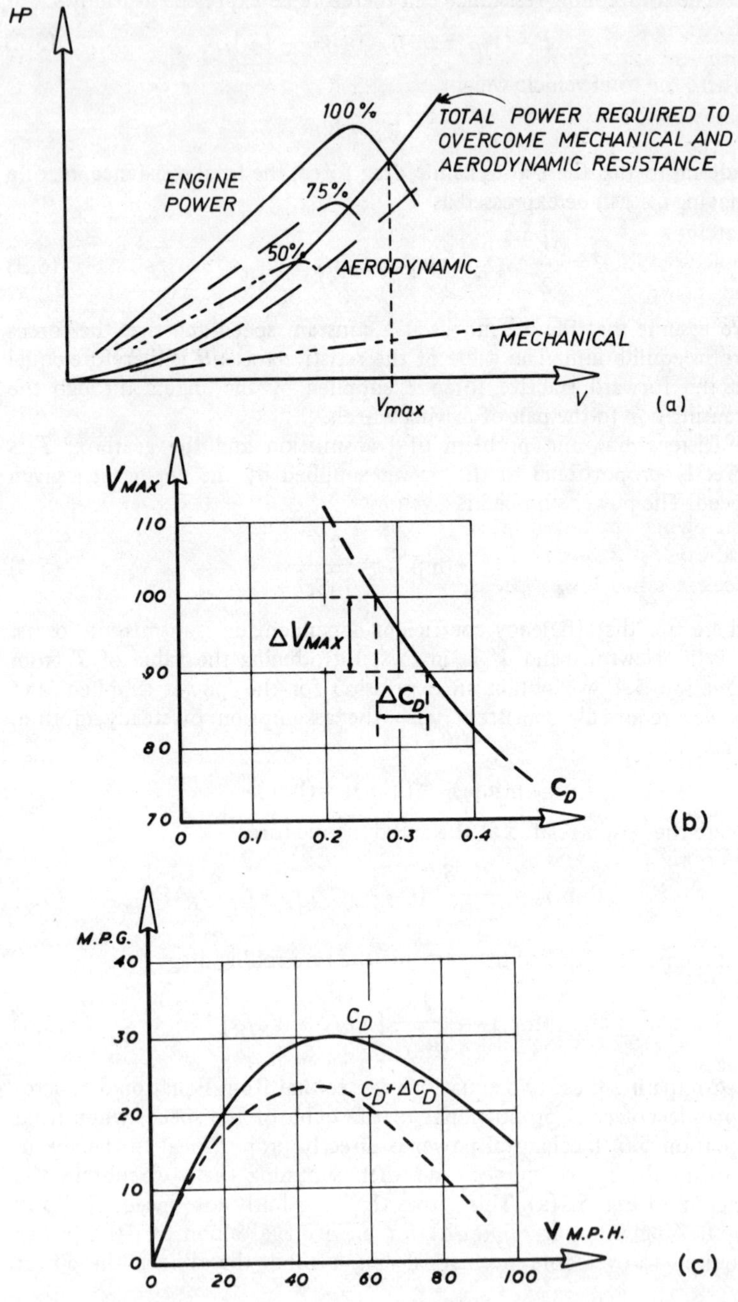
HP
100%
TOTAL POWER REQUIRED TO
OVERCOME MECHANICAL AND
AERODYNAMIC RESISTANCE
ENGINE
POWER
75%
50%
AERODYNAMIC
MECHANICAL
V
Vmax
(a)
VMAX
110
100
90
80
70
ΔVMAX
ΔCD
CD
0
0.1
0.2
0.3
0.4
(b)
M.P.G.
40
30
20
10
0
CD
CD+ΔCD
V M.P.H.
0
20
40
60
80
100
(c)

required to overcome the aerodynamic drag. Above this speed the power required to overcome the aerodynamic drag which increases with the cube of the speed, quickly becomes much greater than that required to overcome the mechanical resistance to rolling, which increases almost linearly with speed.

The power supplied and the torque generated by the engine at a given throttle setting depend on engine rotation speed and this is related to the road speed of the vehicle through the gearing. The matching of engine characteristics to vehicle performance and the choice of gearing ratios is a complex task which interacts with the aerodynamic characteristics and affects the maximum speed attainable, the acceleration at a given speed and the time required to reach a given speed or to achieve a given speed increase. In Fig. 5.1(a) the intersection of the curve representing the useful power available from the engine with the curve representing the power required by the vehicle to maintain the motion, gives the condition of equilibrium for a given car speed. In some cases vehicles are designed so that the maximum speed occurs at an engine speed which is slightly greater than that giving maximum power output at full throttle. In this way there is always a useful margin of excess power or tractive force available for acceleration at lower speeds.

5.1 AERODYNAMIC FORCE AND MAXIMUM SPEED

If the engine power output is to be at its maximum at the maximum speed, V_{max}, then the maximum power supplied must equal the power required by the vehicle. Thus from Equations 5.4, 5.5 and 5.6 we find

$$745.7\eta(\text{h.p.})_{Emax} = \tfrac{1}{2}\rho S V_{max}^3\{C_D - (C_{LF}f_F + C_{LR}f_R)\} + V_{max}(W_F f_F + W_R f_R) \qquad (5.7)$$

Assuming for simplicity that the rolling drag coefficient of the front and rear tyres is the same, i.e. that $f_F = f_R = f$, then we have

$$745.7\eta(\text{h.p.})_{Emax} = \tfrac{1}{2}\rho S V_{max}^3(C_D - C_L f) + V_{max} W f$$

This equation is not easy to solve and is best solved graphically as outlined above. An approximate estimate of V_{max} can be obtained if we assume that about 70% of the engine maximum power is available to match the aerodynamic power required at V_{max}, then

Fig. 5.1. (a) Power required and power available plotted as a function of car speed; (b) effect of aerodynamic drag on maximum speed; (c) effect of increasing aerodynamic drag on fuel consumption for different speeds

$$V_{max} \approx 10\left\{\frac{(\text{h.p.})_{Emax}}{\rho S(C_D - C_L f)}\right\}^{1/3}$$

Alternatively if the engine torque at maximum power output, the gearing and the effective radius of the driving wheels are all known, then the forward tractive force supplied by the engine through the transmission to the driving wheels can be determined. From Equation 5.3 we can calculate the maximum speed of the car expressed in terms of the tractive force at maximum engine power $T_{h.p.\,max}$, and the aerodynamic and rolling drag. The maximum speed of the car is then given by

$$V_{max} = \left\{\frac{T_{h.p.\,max} - Wf}{\dfrac{S}{\rho \frac{}{2}(C_D - C_L f)}}\right\}^{1/2} \tag{5.8}$$

where we have assumed for simplicity that the rolling friction coefficient of the front and rear tyres is the same.

Equations 5.7 and 5.8 allow an analysis of the influence of the aerodynamic drag on the car speed. They show that for a given engine power or tractive force and weight, the speed depends not only on the value of the rolling friction coefficient but on the value of the aerodynamic drag coefficient C_D and also on the aerodynamic lift coefficient of the car, C_L. Decreasing the aerodynamic drag coefficient of the car shape, or the sectional reference area S, produces an increase in the maximum obtainable speed of the car, Fig. 5.1(b). In order to take full advantage of the reduction in drag, the gearing of the transmission should also be adjusted slightly to rematch the power supplied by the engine to that required by the car, as is indicated in Fig. 5.2(a).

The effect of aerodynamic lift on car speed is a little more complicated. From Equation 5.7 or 5.8 we may deduce that an increase in aerodynamic lift reduces the rolling resistance and so tends to increase the speed. On the other hand the aerodynamic lift also affects the maximum tractive force which can be generated. This force, due to the friction of the driving tyres on the road, is given by

$$T_{max} = (W_{R,F} - L_{R,F})\mu$$

where R or F denotes rear or front wheel drive and μ is the tyre-road friction coefficient which depends on tyre tread pattern, tread compound and road surface. By increasing the lift we decrease the maximum tractive force which can be generated at the driving wheels and may take this value below that generated by the engine, and hence we may induce wheel spin. Therefore the total effect of aerodynamic lift on car speed is the result of two conflicting influences which have a comparatively small practical effect on the average saloon car. Only for fast sports cars and racing cars does this problem become significant, and then the need to increase tractive forces through aerodynamic negative lift devices is generally dominant. This problem is, however, very important from the steering and road holding point of view, which is discussed in Chapter 8.

5.2 EFFECT OF AERODYNAMIC DRAG ON FUEL CONSUMPTION

The aerodynamic drag of the car has to be overcome by the tractive force supplied by the engine. Thus the drag directly affects the power required and is reflected in the fuel consumption. The fuel consumption at any speed is directly proportional to the power required. The power required to overcome the aerodynamic drag force is a considerable part of the useful work done by the engine. The overall importance of this item can be shown by the following estimate of vehicle economy. Assume that the total number of cars in the U.K. as a round 1 million. The cars travel at an average speed of 35 m.p.h., say, for about 5000 miles per year. If in doing this they develop an average engine power of about 30 b.h.p. and assuming that the average fuel consumption at this speed and power is 25 m.p.g. then the yearly average fuel consumption is approximately 200 million gallons. Assuming that by careful design and styling the average aerodynamic drag coefficient could be reduced by 2% say, the engine power requirement would be reduced by about 0.5%, thus saving about 1 million gallons of fuel per year. Actual drag reductions can be greater than 2%.

The characteristic fuel consumption of an engine is usually established experimentally and expressed as a nominal value for 1 h.p. as C gal/h.p. hr. Multiplying the actual power of the engine by this characteristic value gives the total engine fuel consumption

$$G = \text{h.p.} \times C(\text{gal/hr})$$

In practice, the fuel consumption is given in the form of a ratio, either distance travelled to fuel consumed, i.e. V/G = m.p.g., which is used in the U.K. and many other countries, or its inverse, the fuel consumed in litres per 100 km travelled, which is generally favoured in Continental Europe.

Consider the case when the aerodynamic characteristics are affected by adding a roof rack. We may say that the original aerodynamic drag coefficient C_D is increased by an addition ΔC_D, and the new total drag coefficient is $C_D + \Delta C_D$. We shall now investigate the effect of this drag increase on the fuel consumption of the car.

The car equipped with roof rack would require more engine power for the same speed to overcome the additional aerodynamic drag force. Writing the ratio of the power related to aerodynamic force before and after the addition of the roof rack (using Equation 5.5), we obtain

$$\frac{(\text{h.p.})_{A\ with\ rack}}{(\text{h.p.})_{A\ no\ rack}} = \frac{(C_D + \Delta C_D) - C_{L2}f}{C_D - C_{L1}f}$$

where C_{L1} and C_{L2} are the values of the aerodynamic lift coefficient before and after fitting the roof rack. Assuming that the characteristic fuel consumption of the engine is unchanged by the change in throttle setting needed to generate the additional power at the same speed, then

$$\frac{\text{m.p.g.}_{no\ rack}}{\text{m.p.g.}_{with\ rack}} = \frac{\frac{1}{2}\rho V^2 S(C_D + \Delta C_D - C_{L2}f) + Wf}{\frac{1}{2}\rho V^2 S(C_D - C_{L1}f) + Wf} \tag{5.9}$$

This example shows that any reduction in the aerodynamic drag coefficient due to styling results in a reduction in fuel consumed which is proportional to the square of the vehicle speed for a given journey. Equation 5.9 also indicates that a reduction in vehicle weight or rolling resistance coefficient results in a proportional fuel saving. The effect on fuel consumption of increasing the aerodynamic drag is shown graphically in Fig. 5.1(c).

Following the fuel crisis in 1974, the public at large has become aware of the vulnerability of petroleum fuel supplies and the finite nature of this resource. Governments have become concerned with energy conservation and the contribution which can be made by the transport industry has been recognised. Vehicle manufacturers and operators or owners have quickly seen the immediate financial benefits in either improved sales or reduced operating costs which result from improved fuel economy. The motoring public need to be able to compare the relative fuel economy of different cars produced by various manufacturers, but fuel economy is highly dependent on the speed and type of journey being made. Various governments have specified typical journey cycles, generally an urban cycle with low speeds and intermittent progress and an open road cycle with higher speeds and motorway type conditions represented, against which the car manufacturers can calculate the fuel economy of their various models.

Following the details of such a set of calculations is beyond the scope of this book as a more detailed knowledge of engine performance is required and in particular the detailed variation of specific fuel consumption with changes of engine speed and load. Typical examples of such performance details have been published [97,102] and the concerned reader should consult these papers.

However, the overall complexity of the detailed problem does not stop us examining the relative importance of the various uses to which the engine work is being put, and hence the various factors which contribute to fuel consumption in the different standardised economy cycles.

In the urban cycles, which have a low average speed of about 20 m.p.h. (about 30 km/hr), work is done in repeatedly accelerating the mass of the car whilst application of the brakes dissipates the kinetic energy of the vehicle as heat. Work is also being done against the mechanical or rolling resistance P and against the aerodynamic drag D_A which, at the low speeds generally permitted on urban roads, is nearly always less than the rolling resistance. If we accelerate a car of mass M from rest to a steady speed V, then the work done is

$$E_1 = \tfrac{1}{2}MV^2$$

The work done against the rolling resistance whilst travelling a distance s is

$$E_2 = Ps = Wfs$$

from Equation 5.2 on neglecting aerodynamic lift.

The work done against the aerodynamic drag whilst travelling a distance s at a steady speed V is

$$E_3 = D_A s = \tfrac{1}{2}\rho V^2 S C_D s$$

Now consider a car of mass 750 kg making a journey of say 15 miles (24 km) length during which the maximum steady driving speed is 30 m.p.h. (48 km/hr) and during which we make 30 stops. Assuming a rolling friction coefficient of 0.02 and a drag area of 7 ft^2 (0.65 m^2), then we find that the ratio

$$E_1 : E_2 : E_3 = 1.2 : 2.0 : 1.0$$

Thus roughly one quarter of the useful engine work would be done in overcoming aerodynamic drag, whilst three quarters of the useful work would be done against forces which were directly proportional to the mass of the car. This shows clearly the great importance of reducing the weight of a vehicle when considering fuel consumption on an urban cycle.

If we now consider a journey of equal distance using the same car but driving at a maximum steady speed of 60 m.p.h. (96 km/hr) and during which we make two stops. Then we find that E_2 is unchanged and so we write the ratio

$$E_1 : E_2 : E_3 = 0.3 : 2.0 : 3.9$$

Thus we see that on an open-road cycle about two thirds or more of the useful work is being done against aerodynamic drag and it is in this type of cycle or of motoring that reduction of drag-area is most beneficial.

5.3. EFFECT OF DRAG ON ACCELERATION

Car manufacturers quote figures giving times to reach certain speeds. This characteristic is not only of prime importance in the case of racing cars but also in day to day situations on the road such as overtaking. In practical motoring the aerodynamic forces acting on a car greatly affect its acceleration capability and these factors must be fully understood.

At a given throttle opening the torque developed by an engine is very nearly constant over a wide range of engine speed. Hence for given

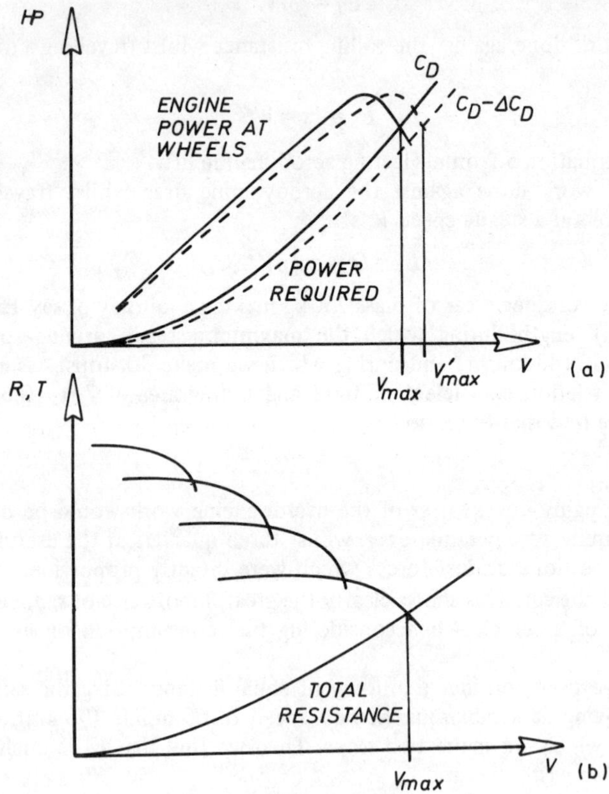

Fig. 5.2. (a) Effect of drag reduction and gearing change on maximum speed of a car; (b) variation with road speed of the tractive force in different gears, and the total resistance of a car

gearing in the transmission, the tractive force T developed at the driving wheels is nearly constant over a range of road speed. Changing the gearing at given road speed will change the tractive force T without changing the engine torque significantly.

Writing the acceleration at the time derivative of speed V, i.e. as $\mathrm{d}V/\mathrm{d}t$ we can relate acceleration of a car of mass M kg to the excess of tractive force over resistive force in level conditions, using Equation 5.3, as

$$\frac{\mathrm{d}V}{\mathrm{d}t} = \frac{T}{M} - \rho \frac{V^2 S}{2M}(C_D - C_L f) - gf \tag{5.10}$$

This is best solved from graphs showing together the tractive force in different gears and the aerodynamic and rolling resistive forces varying with speed, Fig. 5.2(b). The right hand side of Equation 5.10 can then be determined at any road speed and vehicle weight, and the resulting acceleration calculated. On examining Equation 5.10 we see that the greater the mass of the vehicle the smaller is its acceleration for a given tractive force. In fact the acceleration is dependent on T/M which relates to the vehicle power to weight ratio, and best values will be achieved at full throttle. Acceleration is reduced by the aerodynamic drag which increases with speed squared, see Fig. 5.3(a), so if we increase vehicle drag coefficient from C_D to $C_D + \Delta C_D$, the acceleration at speed V is changed by an amount

$$\Delta\left(\frac{\mathrm{d}V}{\mathrm{d}t}\right) = -\rho \frac{V^2 S}{2M} \Delta C_D$$

The time to accelerate from speed V_1 to V_2 can be found if we know the variation of acceleration with speed from Equation 5.10. Since the time t is related to the reciprocal of the acceleration, we can determine t by plotting a graph of $\{\mathrm{d}V/\mathrm{d}t\}^{-1}$ against speed and measuring the area enclosed under the graph between speeds V_1 and V_2, Fig. 5.3(b). Detailed analysis of this problem shows that the time taken increases with the weight of the vehicle and that an increase of drag coefficient by ΔC_D results in an increase in the time taken of approximately

$$\Delta t = t \frac{\Delta C_D}{2C_D}$$

In a number of situations which can arise in both road and competition driving, the driver has a limited distance in which he can accelerate. It can then be useful to relate change of speed with distance required,

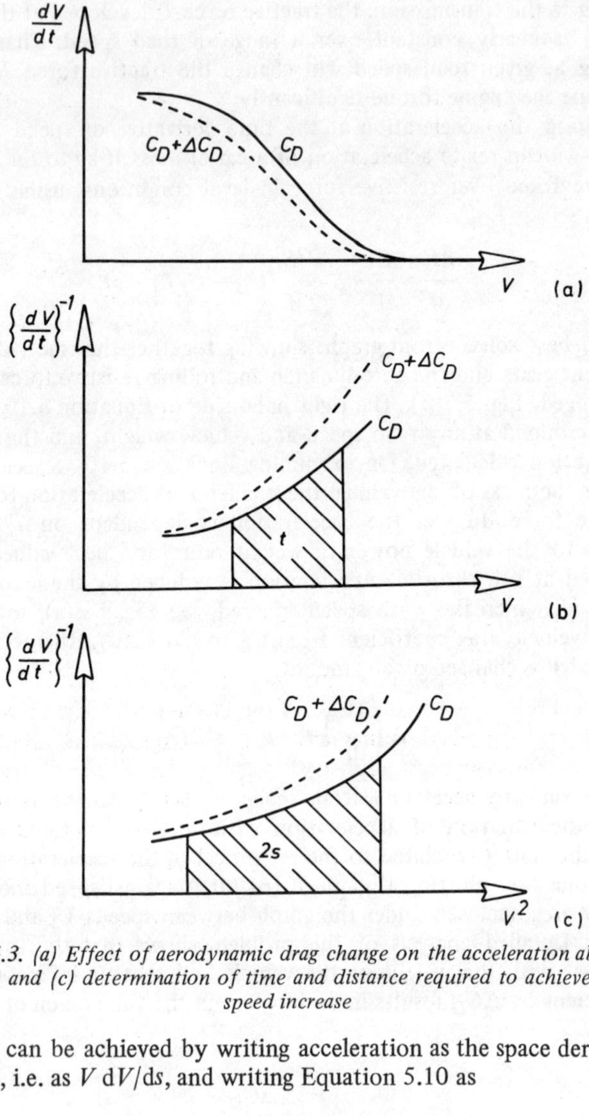

Fig. 5.3. (a) Effect of aerodynamic drag change on the acceleration ability of a car; (b) and (c) determination of time and distance required to achieve a given speed increase

and this can be achieved by writing acceleration as the space derivative of speed, i.e. as $V\,\mathrm{d}V/\mathrm{d}s$, and writing Equation 5.10 as

$$V\frac{\mathrm{d}V}{\mathrm{d}s} = \frac{T}{M} - \rho\,\frac{V^2 S}{2M}(C_D - C_L f) - gf$$

Since time and distance covered are related through speed, we can find the distance covered, s, by plotting a graph of $\{\mathrm{d}V/\mathrm{d}t\}^{-1}$ against speed squared and measuring the area enclosed under the graph, Fig. 5.3(c). However, if we assume that tractive force T is constant and we denote by V_m the characteristic speed of the vehicle at which the acceleration is zero at this tractive force, we can solve the above equation to give the distance

$$s = \frac{2.303M}{\rho S(C_D - C_L f)} \log \frac{(1 - V_1^2/V_m^2)}{(1 - V_2^2/V_m^2)} \tag{5.11}$$

Careful examination of all the terms in this equation shows that an increase in vehicle drag coefficient increases the distance required to achieve a given speed increase or reduces the speed increase which can be attained in a given distance.

The ability to accelerate quickly and efficiently is an important characteristic directly related to safety in traffic. We must, therefore, come to the conclusion that aerodynamic styling has an important role to play in this matter.

5.4 EFFECT OF DOWNFORCE ON CORNERING

One of the main benefits of the aerodynamic downforce or negative lift force developed on racing cars by the use of inverted wings and ground effect body shaping has been to improve road holding. In particular this has been applied to increase the speed of cars through curves, enabling them to exit at a higher speed and then to accelerate over an often-constrained distance to reach a higher speed than would otherwise be attained. Both these effects help to reduce lap times on racing circuits.

For a racing car tyre, on combining the driving thrust, rolling resistance and side force, the resultant force on the road acting in the plane of the road, is proportional to the download on the wheel, as indicated in Fig. 5.4(a). Allowing for the driving thrust and rolling resistance allows us to estimate the maximum side force Y_T on a racing tyre supporting a load W by

$$Y_T = \mu W$$

where μ is the effective adhesion factor of the tyre which depends on the tread compound, temperature and inflation pressure.

Consider a car of mass M in a steady turn of radius R at speed V on a level unbanked track in still air, as shown in Fig. 5.4(b). Applying Newton's law of motion in the inward radial direction gives

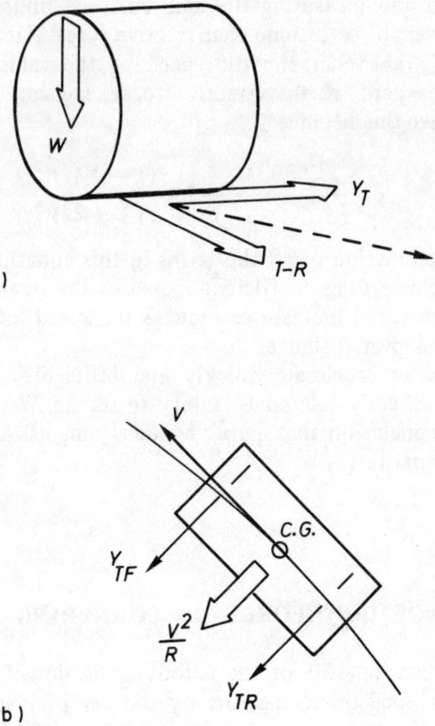

Fig. 5.4 (a) Loads on a tyre; (b) tyre forces on a car

$$M\frac{V^2}{R} = Y_{TF} + Y_{TR}$$
$$= \mu_F W_F + \mu_R W_R = \mu W$$

where for simplicity we have assumed that $\mu_F = \mu_R = \mu$ and that W is the total download. The total download consists of the total weight of the vehicle less the aerodynamic lift L_A which we can write in coefficient form using Equation 2.3(b). Thus we have

$$M\frac{V^2}{R} = \mu\{Mg - \tfrac{1}{2}\rho V^2 SC_L\} \tag{5.12}$$

and from this equation we can determine the speed in the curve

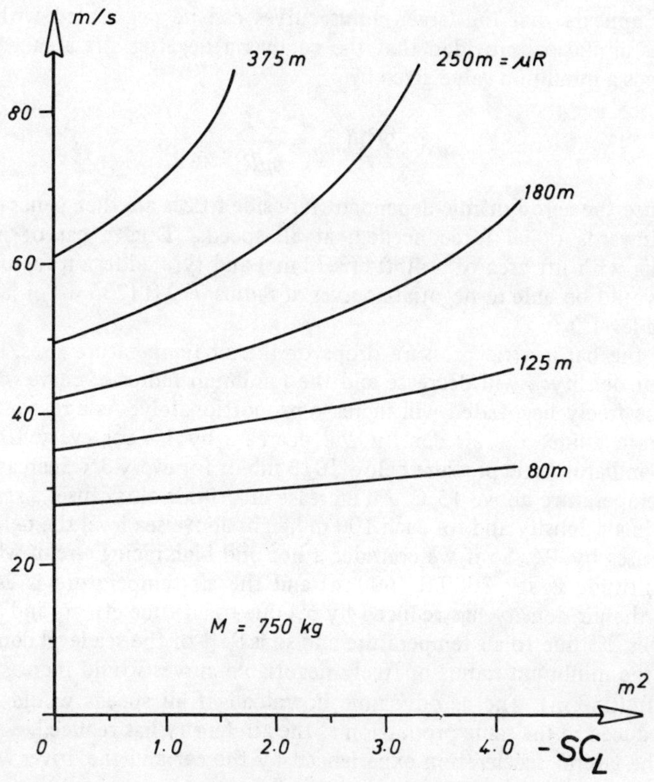

Fig. 5.5. Variation of maximum speed through a corner of given radius with lift area of a racing car

$$V^2 = \frac{Mg}{\dfrac{M}{\mu R} + \dfrac{\rho}{2} SC_L} \tag{5.13}$$

remembering that C_L will be negative for downforce.

For a given car mass, the dependence of the maximum steady road speed through a curve can be determined against the product of reference area and lift coefficient (i.e. the lift-area) for various values of the radius of the curve and the tyre adhesion parameter. Such information is presented in the set of graphs shown in Fig. 5.5 which have been calculated for a car with mass 750 kg; curves can be negotiated at all speeds below those delimited by the particular μR value, and to the left of the negative lift area of the car, which is assumed to be known.

It appears that the larger-radius curves can be negotiated without speed limitation provided that the car has a negative lift area which exceeds a minimum value given by

$$(-SC_L)_{min} = \frac{2M}{\rho\mu R}$$

because the aerodynamic-dependent tyre side forces are then generating the inwards radial force needed, at all speeds. Thus a car of mass 750 kg with lift area of -40 ft^2 (3.72 m^2) and tyre adhesion factor of 1.4 would be able to negotiate curves of radius 770 ft (235 m) or larger at sea level.

If the barometric pressure drops or the air temperature rises, then the air density ρ will decrease and the minimum radius of curve which can be freely negotiated will increase proportionately. As a reasonably accurate guide, the air density will decrease by 1% for every 10 mb drop in barometric pressure below 1013 mb or for every 3°C increase in air temperature above 15°C. An increase in altitude also causes a reduction in air density and for each 100 m height above sea level the density decreases by 1%. So if we consider a hot and high racing circuit where the altitude is say 2000 ft (600 m) and the air temperature is 24°C, then the air density has reduced by 6% due to altitude effects and by a further 3% due to air temperature and so is 91% of the sea level density and the minimum radius of freely negotiable curves would increase to 846 ft (258 m). The aerodynamic download at all speeds would also be reduced in the same proportion as the air density has reduced.

The lateral acceleration experienced by the car and the driver when negotiating curves is increased due to the aerodynamic downforce. From Equation 5.12, expressing the lateral acceleration in 'gravity' multiples gives

$$\frac{V^2}{Rg} = \mu \left\{ 1 - \frac{\rho V^2 SC_L}{2Mg} \right\} \tag{5.14}$$

which shows that the lateral acceleration increases directly with the ratio of aerodynamic download to vehicle weight, and it is no surprise that drivers are experiencing lateral accelerations in excess of 3*g*.

5.5 EFFECT OF DOWNFORCE AND DRAG ON BRAKING

The braking force generated between a tyre and the road surface is proportional to the total load W carried on the tyre and the adhesion factor μ of the tyre. The total download consists of the total weight of the vehicle, Mg, less the aerodynamic lift L_A which we can write in

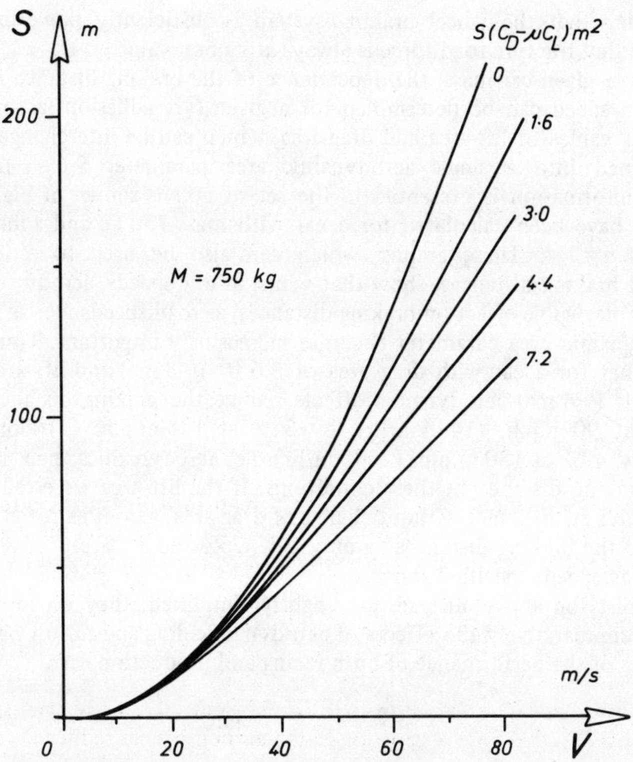

Fig. 5.6. Effect of aerodynamic area in reducing braking distances

coefficient form using Equation 2.3(b). The vehicle is being retarded at all times by the aerodynamic drag D_A which also has a coefficient form given by Equation 2.3(a). For simplicity we will assume that the adhesion factors of the front and rear tyres are the same and then we can write the retarding force R acting on a vehicle as

$$R = \mu Mg + \tfrac{1}{2}\rho V^2 S(C_D - \mu C_L) \tag{5.15}$$

where C_L will be negative for downforce.

Writing the deceleration as $-V\,dV/ds$ then enables us to solve this equation, and we find that the braking distance s required to come to rest from speed V is given by

$$s = \frac{2.303M}{\rho S(C_D - \mu C_L)} \log\left\{1 + \frac{\rho S(C_D - \mu C_L)}{2\mu Mg} V^2\right\} \tag{5.16}$$

assuming that the wheel braking system is sufficiently powerful to ensure that the tyre-road force is always at its best value.

For a given car mass, the dependence of the braking distance from a given speed can be determined for a given tyre adhesion factor for various values of lift-area and drag-area, which can be interchanged or combined into a single aerodynamic area parameter $S(C_D - \mu C_L)$. Such information is presented in the set of graphs shown in Fig. 5.6 which have been calculated for a car with mass 750 kg and adhesion factor $\mu = 1.4$. These graphs, which can also be used to estimate partial braking distances, show that whilst at low speeds, aerodynamics have a negligible effect on braking distances, as road speeds increase the aerodynamic area parameter becomes increasingly important. Thus we find that for a car with drag area of 8.6 ft^2 (0.8 m^2) and lift area of -43 ft^3 (-4 m^2) aerodynamic effects reduce the braking distance by 22% at 90 m.p.h. (144 km/hr), by 37% at 135 m.p.h. (216 km/hr) and by 49% at 180 m.p.h. (288 km/hr), i.e. aerodynamics then halves the braking distance in these conditions. If the lift area were reduced to -21.5 ft^2 (-2 m^2) without change of drag area, aerodynamic effects reduce the braking distances by only 14%, 26% and 37% respectively at the road speeds specified above.

Whilst the above analyses are slightly simplified, they do serve to show comparatively the effects of aerodynamic drag and lift on various aspects of the performance of both racing and production cars.

Chapter 6

AERODYNAMICS AND STYLING

The styling of a car is the result of a unique and creative cooperation between designer, aerodynamicist, engineer and manufacturer. This creative relationship is arranged and organised in various ways by different car manufacturers but the main principles followed in this process of styling are basically the same in each case.

In the creative process of styling the designer has to consider many aspects, such as structural design and general handling and safety, but above all he has to choose a shape which is comparatively easy to produce. Aerodynamic considerations are only one of many, but they are all directly related to car performance, economy and handling characteristics, which in turn strongly influence the safety of the car. The choice of the car shape is, therefore, one of fundamental importance.

This chapter should be regarded as an introduction to the process of styling. It is based more on what styling should be than is actually the case at present but it may, however, give a certain amount of insight into the problem of how a particular shape is chosen.

The design of such a complex machine as an automobile requires the application of the art of compromise from the designer. In this case it is often a compromise between the demands of production cost, structural requirements, or simply driver and passenger comfort. Among the many, often conflicting, requirements is the problem of the choice of an external shape in order to obtain suitable aerodynamic characteristics, which in turn affect performance and safety.

A new car project is usually the product of a long and elaborate design study by a research team. This team establishes what basic unit structure to adopt, chooses the most suitable materials and production methods and tries to achieve an attractive appearance as well as performance. The team's work also reflects the general philosophy of design such as durability, quality and finish, while keeping in mind the need for ease of maintainance and good repair characteristics.

When the basic technical requirements are established the exterior shape of the car has to be visualised. This shaping is usually carried out by the styling department and occurs at a very early stage of the prototype development. The car is then still in a formative stage and may be subjected to quite major alterations. This stage of styling is mainly influenced by general production requirements, and is usually based on previous experience with similar types. The problem of how big a steel sheet pressing the production method economically allows, or how many major parts have to be assembled on the production line, has to be solved. The exterior shape of the car is usually determined by the current fashion prevailing at the time. In practice the form is usually established using china clay sculpture.

6.1 THE SCALE MODEL AND AERODYNAMIC TESTS

At this stage a simple aerodynamic model of the car project is made, 1/5th or 1/4th of the full scale size. This model is tested in the wind tunnel to obtain preliminary aerodynamic characteristics such as the aerodynamic drag, lift and side-force coefficients. These coefficients are then compared with those of other cars, especially those in a similar category. If these characteristics are found to be not particularly good, that is deviating too much from the desired values, the aerodynamicist is given the job of improving them by making the appropriate changes to the shape.

It is important to be aware at the very beginning of all the basic aerodynamic principles involved to avoid making a fundamental error. In other words the logical connection between cause and effect, that is between the choice of any particular car shape (or part of it) and the resulting effect on the aerodynamic characteristics, must be known in advance. Basic mistakes in shaping the model are sometimes difficult to overcome by a simple modification and a totally different approach is then necessary.

Results of wind tunnel measurements are passed to a team of designers to give them the data required for preliminary performance calculation. At this stage certain design requirements may have to be fulfilled. Often this concerns things like the inlet and outlet of air ducts serving either the engine cooling system or the cabin ventilation system.

To give designers clear-cut answers on where to put the inlet to the radiator, or the ventilation grills, or their size, a more detailed investigation of the pressure distribution on the surface of a model is often required. At this stage a flow visualisation method is frequently used which is the quickest quantitative method of studying the airflow pattern.

Concurrently with the scale model used in the wind tunnel tests the design engineers further develop the scale model, made from hard clay, which helps in the development of the structural design, and at the same time acts as a check on the shape changes. Some designers build full-sized mock make-ups, made of wood, in which a study of the passenger compartment size and proportions, the arrangement of the instrumentation and sometimes even the ventilation system, are tested.

All this preliminary work results in the creation of a design in which the various requirements are coordinated as best as possible. Final modifications to the model shape are made at this stage. The finalised scale model is then used in the wind tunnel experimental tests, the results of which, i.e. the basic aerodynamic characteristics, are fed-back to the design office. This aerodynamic data is now used to calculate the performance, and to estimate the road holding and stability characteristics of the car project.

In this briefly sketched process of a prototype design the aerodynamic styling plays a gradually increasing role. The importance of the creative stage, when the car shape is being adapted to improve the aerodynamic characteristics, varies according to the particular type of car. It would be more important in the case of a fast car than in the case of a slow one built to some economic specification but it is important in any case, especially from the point of view of safety. The styling of the body has two main requirements, firstly to reduce the aerodynamic drag, and secondly to reduce the aerodynamic lift. Another aerodynamic characteristic which is of importance is the rate of change of the aerodynamic side-force with the yaw angle. The body shape of a road vehicle is already a compromise between the demands of the mechanical design and the need for driver and passenger comfort and safety. All this, it should not be forgotten, is under the pressure of commercial considerations, expressed by the production engineers on the one side and the market research departments on the other.

6.2 IDEAL (MINIMUM DRAG) SHAPE AS A STYLING CRITERION

The car is shaped around its occupants, and its size is determined by the number of passengers and seating arrangement. Consider the usual four-seater; the styling problem is to wrap-up a good aerodynamic shape around the occupants. A fundamental dimension is the 'maximum cross-sectional area' S. This area is chosen to accomodate the occupants comfortably and may be considered as the characteristic dimension. As the car is a three-dimensional object its volume could be shaped in many different ways and is a compromise dictated by structural and mechanical requirements, and the maximum cross-sectional area. This shape should have optimal aerodynamic characteristics. Comparative values of the drag coefficients of different types of cars corresponding to typical values of the length-to-height ratio are given in Fig. 6.1. For a given value of S there is, theoretically, an optimal shape which when wrapped around this cross-sectional area gives the minimum possible aerodynamic drag.

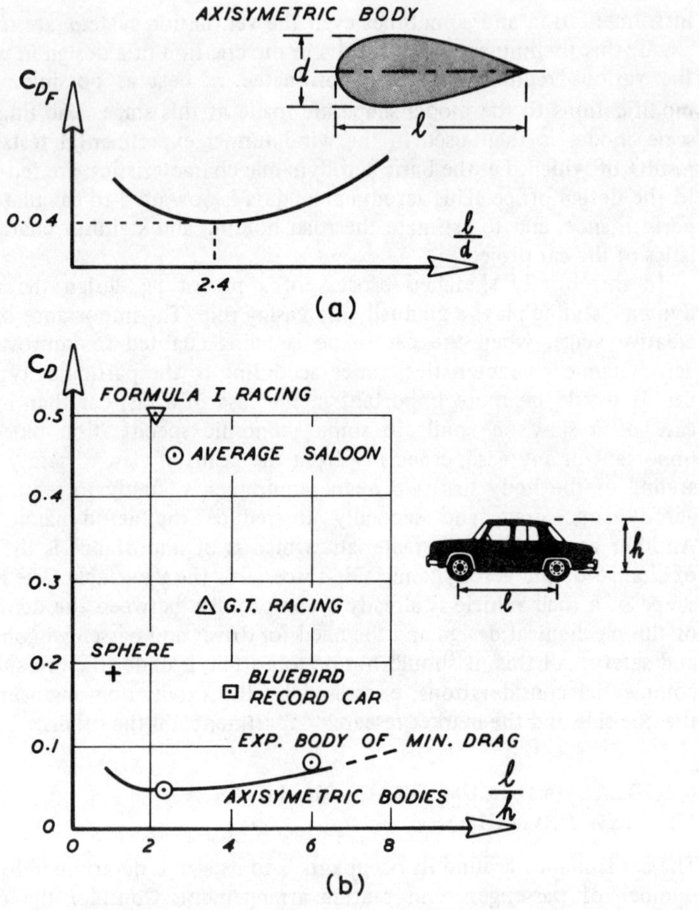

Fig. 6.1. (a) Effect of the 'slimness' of an axisymmetric body on the aerodynamic drag coefficient. (b) Effect of the length-to-height ratio on the drag coefficient for different types of car

Attempts to obtain such an idealistic shape in simulated airstream conditions and in the proximity of the ground have been made, among others, by the Author [48]. The ideal shape in free flight, i.e. in an unrestricted airflow away from the ground, is an axisymmetric body with a circular maximum cross-section, Fig. 6.1(a). Theory and experiment have shown that the aerodynamic drag of this shape is comparatively high when the length is small, that it falls to a minimum of 0.04 when the length to diameter is about 2.4 and then starts to increase again as the length increases. This minimum value occurs when the relationship between the normal pressure distribution (form drag) and that due to skin friction (which increase more rapidly with length than does pressure drag) is optimum.

This axisymmetric body is not an ideal shape for motion near the ground. The presence of the ground introduces an asymmetry of the flow pattern and modification of the shape is then required to minimise ground interference. It appears that it is difficult to develop a mathematical model, i.e. a mathematical formula which transforms a three-dimensional body moving in free space to a representative surface of the body moving near the ground. Even if this were achieved it would still be a solution representing the airflow around a mathematically ideal body, but in a fluid devoid of viscosity. We know from previous discussions that the motion near the ground is greatly affected by viscosity which plays a dominant role.

Many scientific workers have tried and failed, as far as it is known, to develop such an ideal shape by mathematical analysis. One possible solution is to create an arbitrary shape which represents a good approximation to the hypothetically 'ideal shape'. This approach has been taken by other investigators [29,49] and by the Author. Efforts by car designers who have created minimum drag bodies in their attempts to reach a new land speed record can also be included in this category.

The approach made by the Author involved taking as the maximum cross-sectional area a form comprised of two half-ellipses joined along their (equal) major axis, (Fig. 6.2(b)) the lower half-ellipse having a smaller minor axis than the upper one. The major axis was made to vary with distance from the nose according to the geometry of a low-drag symmetric aerofoil. The minor axes of the lower and upper half-ellipses were chosen to accord with the geometry of a cambered aerofoil contour with good aeronautical characteristics. The elliptical cross-sections produced a 'whale-shaped' body as the transformations were made in a plane crossing the main axis at right angles. The resultant shape is shown in Fig. 6.2(a). Statistically the dimensions of an average road vehicle show the maximum cross-sectional area to be situated at a point approximately 40% of the total length, as measured from the nose. Thus the base of the vertical plane contour was made from a cambered NACA aerofoil of 16% thickness [48], while a NACA symmetric aerofoil of 30% thickness formed the planform contour. The height measured from the chord line at any station x, becomes the minor axis of the half-ellipses b_u or b_l, while the major axis a, which is common to both half-ellipses (upper and lower) is half the width of the symmetric aerofoil.

Such a shape was actually made and tested in a wind tunnel in the presence of the ground represented by a rigid board. The drag

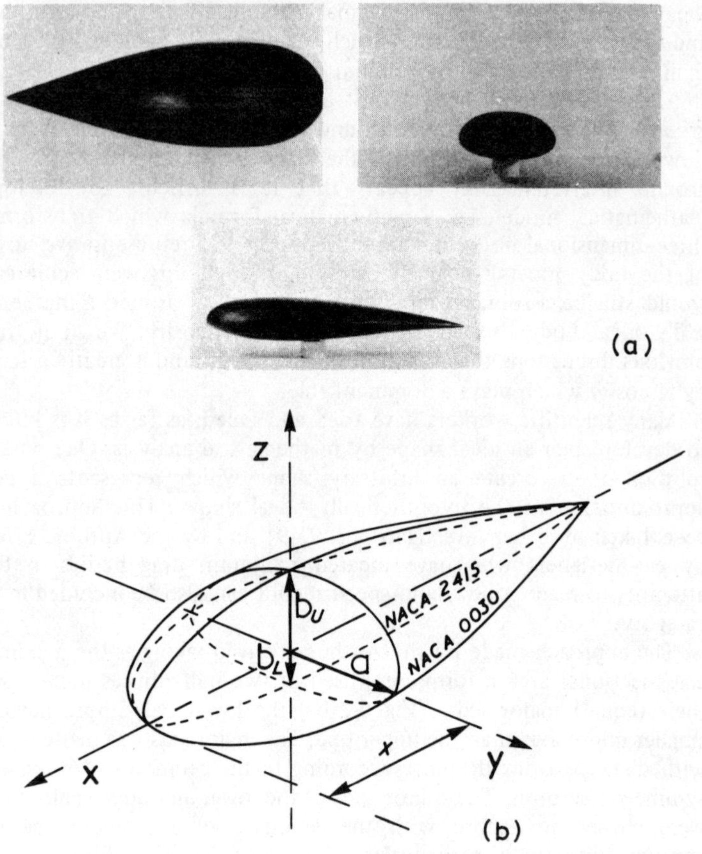

Fig. 6.2. (a) Body of minimum drag (ideal shape); (b) geometry of body of minimum drag

coefficient of this body was found to be of the order $C_D = 0.07$. Better simulation of the ground by the 'moving ground' facility [50,51,52] could introduce a further correction to the results. Other experimental investigations of scale models of fully streamlined bodies, such as the world land speed record car 'Bluebird', gave the minimum drag coefficient as $C_D = 0.16$ (see Fig. 6.1(a)). In comparison with these values, the modern well-streamlined racing car or sports car has a $C_D = 0.2$ to 0.3. The difference in value of these drag coefficients is due mostly to secondary aerodynamic causes originating in deviations from the ideal shape, usually made necessary by the practical considerations of the car design, such as a protruding passenger compartment. To illustrate the drag analysis let us re-arrange the drag equation in the following way: the total aerodynamic drag

$$D_A = D_{ideal} + \Sigma\Delta D \tag{6.1}$$

and the drag coefficient

$$C_D = C_{D\ ideal} + \Sigma\Delta C_D$$

$\Sigma\Delta C_D$ represents the sum of all the additional drag components caused by deviation from the ideal shape. An important deviation from this ideal shape, i.e. bobtailing was first mentioned in Chapter 3 when discussing the general pattern of the airflow and the flow separation. The ideal shape, as shown in Fig. 6.2, is unnecessarily long for road use due to the fact that the airflow would separate from the body well before the end of the tail. The rear part of the 'tail' is therefore not really required from an aerodynamic point of view, hence the process of bobtailing.

The next deviation from the ideal shape is unavoidable from a mechanical point of view. A car moves by rolling on wheels; the interference between rotating wheels constitutes perhaps the most important deviation from the ideal. Another necessary addition to the body shape is the passenger compartment. This constitutes a further important source of interference as it has to provide a forward field of vision which means that it has to stand out from the main streamlined body forming a step into which the windscreen is fitted. In addition to these necessary variations to the ideal shape, we have also to consider the problem of the internal flow discussed in Chapter 3.

Comparing the drag coefficient of the ideal car shape with that of a typical car, we find that the sum of the drag increments C_D amounts approximately to 60% of the total drag, Fig. 6.3(a). Statistical values of the average drag coefficients for typical saloon cars, streamlined racing cars and heavy goods vehicles are expressed in the terms of the ratio $C_{D\ ideal}/C_D$, shown in Fig. 6.3(b).

The stylist, approaching the problem of creating a good aerodynamic shape, may adopt a number of practical methods. However, it seems that by bearing in mind the concept of the 'ideal shape' and knowing just how far one can deviate from it, the stylist will achieve better results than by some 'rule of thumb' method. Practical possibilities for reducing the aerodynamic drag are obviously within the 60% discrepancy between the drag of present day cars and the 'ideal' situation, Fig. 6.3(c). Listing the common sources of drag as an aid when considering the possible ways of reducing drag by styling, we have

(1) Form drag, caused by the horizontal component of the integrated pressure distribution.

(2) Induced drag, caused by the presence of vortices, and consequently aerodynamic lift (sometimes called the 'lift-induced' drag).

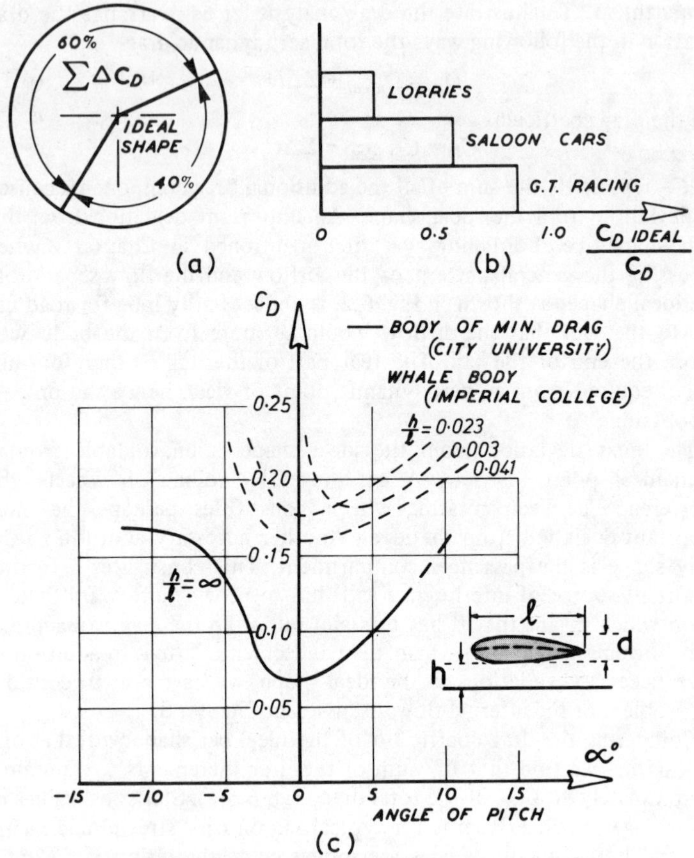

Fig. 6.3. (a) Division of the total drag coefficient into the drag of an ideal body and drag due to the deviation from the ideal shape; (b) ratio of ideal body drag to total drag for selected types of road vehicles; (c) drag coefficient of the idealised body as a function of pitching angle

(3) Interference drag, caused by features on the surface affecting the local airflow.

(4) Internal flow drag, caused by taking a quantity of air from the external airflow and inducing it to flow through a system of internal ducts.

Other main sources of aerodynamic drag, i.e. rotating wheels and the underside flow have been discussed separately. 'Form drag' can be measured theoretically by integrating the horizontal air pressure components which act in a direction opposing the main airstream over the car surface. This method is not very practical but it helps, by visualising the process, to find possible ways of reducing form drag. Any shape at the front which either causes a sudden local acceleration or deceleration of the flow, or a sudden change of flow direction, will add to the form drag. As the upper part of the car includes the windscreen and passenger compartment the airflow has to move over an additional 'obstacle' which accelerates it and directs it sideways. Styling should therefore avoid sudden changes in the actual cross-sectional areas of a car since such changes cause an additional disturbance which increases the momentum losses and thus increases the air resistance. In the front of the car the cross-sectional area increases with lengthwise distance, while downstream of the maximum cross-sectional area it decreases with distance, Fig. 6.4. The dS/dx derivative should, however, always be small.

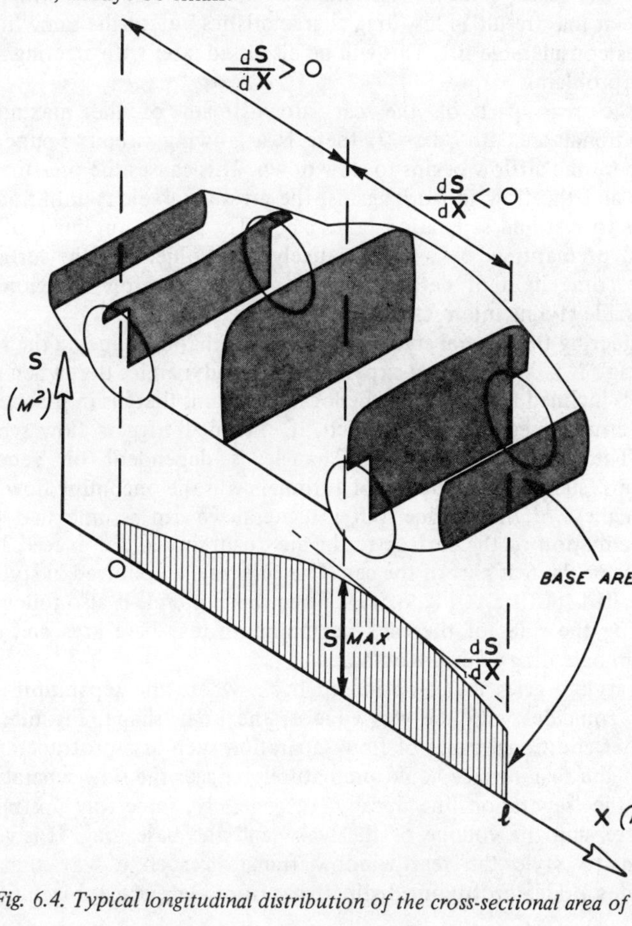

Fig. 6.4. Typical longitudinal distribution of the cross-sectional area of a car

In order to decrease the aerodynamic drag any features which lead to a channeling of the airflow should be avoided. Any shape which helps the local airstream to 'sluice' over the surface causes a local increase in velocity. Such a local jet of air is often crosswise to the main flow pattern and moving faster than the surrounding air it causes turbulence, and thus momentum losses, which add to the air resistance. Local airspeed augmentation is sometimes caused by positioning elevated headlamps at the front of the bonnet.

Air forced to flow between the underside and the ground is subject to a great deal of turbulence. Problems related to the proper styling of underside panels have been discussed in Chapters 3 and 4.

The styling of the windscreen-bonnet configuration should be a compromise. The smooth, streamlined shape of a well wrapped windscreen may result in low drag characteristics but, at the same time, it creates considerable lift. This will be discussed later with steering and stability problems.

At the rear part of the car, downstream of the maximum cross-sectional area, ($\mathrm{d}S/\mathrm{d}x = 0$) there is a growing viscous boundary layer where the airflow begins to slow down. This causes the pressure to increase and the flow to work against the pressure gradient until finally it comes to rest and separation takes place. The separation line is often triggered prematurely by a comparatively small object on the surface, possibly some item of detailed styling, and is therefore a region of considerable styling interest.

Considering the general styling problem, i.e. that of reducing the rear 'base drag', it is known from experimental aerodynamics that when the surface is inclined negatively to the local airstream, there is in practice a certain critical negative angle which, if exceeded triggers flow separation. The value of this critical angle is dependent on general conditions, such as the amount of turbulence in the oncoming flow, or the roughness of the surface, but in general we can assume that the local declination of the surface to the airstream should not exceed 3 to 5 degrees at the rear part of the car. This principle is followed in styling the so-called 'fastback', Fig. 6.5(a). When this principle is also followed in shaping the sides of the car, the minimum rear base area and the minimum base drag value are obtained.

This styling gives us an optimised back, where the separation line roughly coincides with the rear edge of the body shape, Fig. 6.5(b). Any other additional cause of flow separation such as a protruberance fixed on the rear surface could prematurely trigger the flow separation moving the 'separation line' locally, or generally, some way upstream thus increasing the volume of the wake and the base drag. It is very common to style the rear window frame in such a way that it constitutes a ridge protruding distinctively from the surface, Fig. 6.6(a).

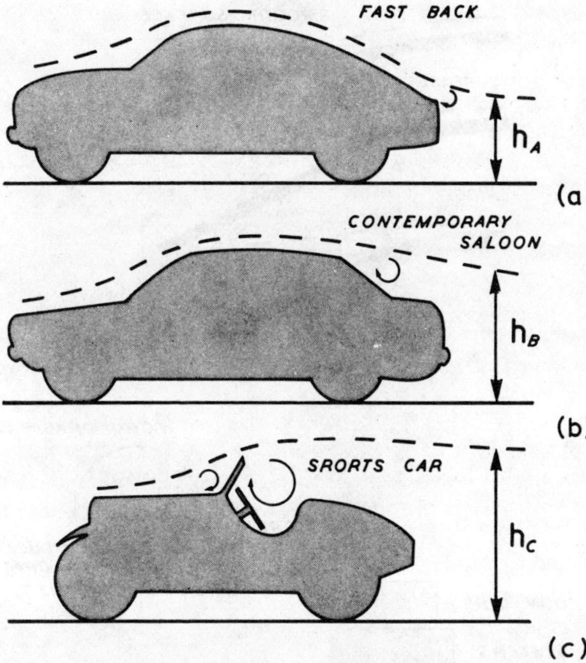

Fig. 6.5. Typical wake pattern of (a) a fast-back car, (b) a contemporary saloon, and (c) an open sports car

Such a ridge placed on a generally declined surface often triggers the separation with the result that the wake starts there, and not further downstream on the edge of the elaborately styled fastback. Thus careful general styling is made inoperative by inattention to points of detail. There are only a few car manufacturers who make windows flush with the surface, and the window frames properly recessed. This is only a typical example and there are many other cases where points of styling detail are equally important.

The same problem exists at the wheel arches, and corresponding cavities, when air flows from them across the main airstream (see Chapter 3, Section 4, dealing with the effect of wheel rotation). This is a problem which is difficult to solve in practice–because of the wheel rotation and the interaction with the flow underneath air tends to flow across the wheel arches from both front and rear. This usually causes separation bubbles which form downstream of the arches. After a wet or dusty run the areas covered by dirt are seen to be just behind the arches where the local turbulence assists the dirt to settle on the car surface. It is not simply a matter of cleaning but a

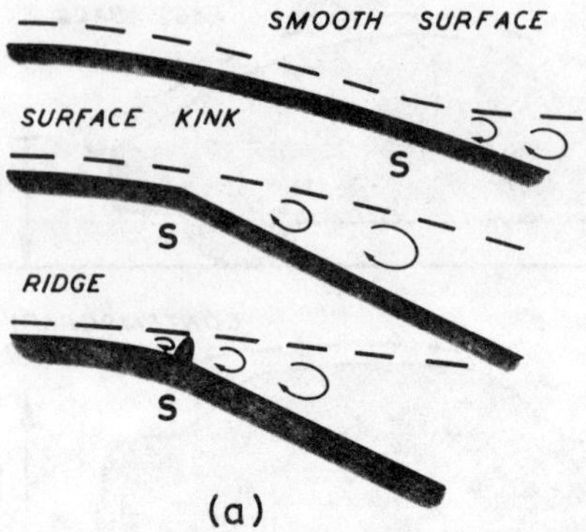

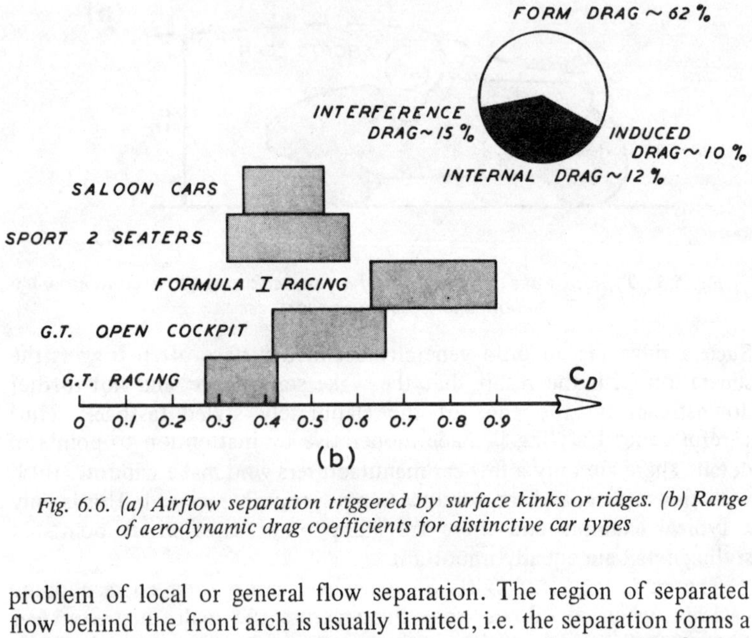

Fig. 6.6. (a) Airflow separation triggered by surface kinks or ridges. (b) Range of aerodynamic drag coefficients for distinctive car types

problem of local or general flow separation. The region of separated flow behind the front arch is usually limited, i.e. the separation forms a bubble and the flow re-attaches itself again to the car side further downstream. The efflux of air from the rear wheel arch causes a separation of the flow which is usually linked with the rear wake. It enlarges the wake to cover this rear part of the car surface and thus increases the base drag. The only styling which may be suggested here is to shield the wheel cavities, either totally or, at least partially. The effect of such shielding was mentioned qualitatively in the discussion in Chapter 3 on rotating wheels.

For sports two-seaters, or open sports cars generally, the flow separation occurs on the upper edges of the windscreen. This, and the open compartment behind, causes the wake to extend over the whole of the upper part of the car, practically up to the maximum cross-sectional area line, Fig. 6.5(c). The general principle of styling involved is to try to reduce the rear base drag by reducing the maximum cross-sectional area. This is achieved by making the rear part of the cabin slope gently backwards, and the whole cabin made as low as is practically possible. For an average car, i.e. with a wheelbase ot 7.5 ft and total length of 14 ft, the maximum cross-sectional area S can be thus reduced by about 40%.

6.3 BASE DRAG REDUCTION BY INCREASING THE WAKE PRESSURE

Bearing in mind that the base drag is a result of wake negative pressures integrated over the base area, it has been suggested that by artificially increasing the pressure within the wake, we may reduce the drag. This has been investigated by Sykes [53] and was found feasible. By ejecting a sufficient mass of air, or exhaust gas, at a low flow rate into the wake we can significantly increase the mean static pressure inside the wake. This method was first studied by aerodynamicists investigating the problems of missile drag. Ejecting air at low flow rate and velocity produces negligible thrust due to the small mass momentum, but it does introduce a higher local pressure.

Consider the total volume of air displaced by a car in motion as the product of its maximum cross-sectional area S and its speed V, i.e. SV is a 'measure' value of the air flow rate. Thus the exhaust gases from a 1.6 litre car engine produce a gas flow of order of 0.67% of the total value of SV, the ventilating system may produce 0.167% and the engine cooling system about 1.2%. All these sources of air or gas can therefore produce a combined flow rate into the wake of about 2% of the total air displacement SV.

In the case of a vehicle with an engine of 5 litres capacity the exhaust gases alone can create an injection rate of about 2.1%. As the exhaust gases are ejected from the exhaust pipe at high speed a plenum chamber is required to reduce the momentum. By losing momentum the gas increases its pressure. From this plenum chamber gas or air, or a mixture of both, is ejected through a grid to ensure the correct rate of flow into the wake [53], Fig. 6.7(a) and (b). The diagram shows the experimental results of a reduction of the base pressure coefficient C_{pb} expressed in terms of the ratio h/D, i.e. the pressure coefficient

$$C_{pb} = \frac{p - p_o}{\frac{1}{2}\rho V^2}$$

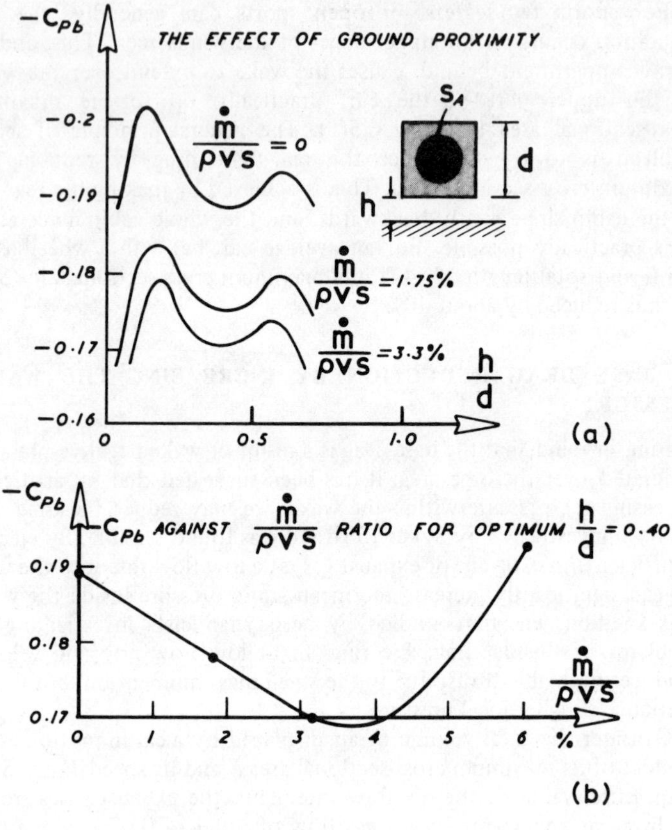

Fig. 6.7. (a) Base drag as a function of the base area and the ground clearance for different gas flow rates into the wake. (b) Relation between base drag and the gas flow rate into the wake (after Sykes)

where p is the local pressure in the wake and p_o is the ambient pressure in the free, undisturbed airstream. Fig. 6.7(b) shows that the C_{pb} curve has a minimum value and by arranging for a gas ejection rate at this value of flow rate the optimum condition for reduction of the base drag is obtained.

Experiments have shown that the effect of the ground proximity is to increase the base drag. Thus low slung cars have a greater base drag than those with large ground clearance, but the effect of the low flow rate gas ejection into such wakes decreases the drag considerably. It is estimated that an approximately 30% reduction of base drag by this method is possible. It seems that by utilising the engine exhaust gases alone, bleeding through a suitable grid at the back of the car, it should be possible to reduce the base drag by up to 10% of its original value.

Fast-car styling developed in Italy seems to indicate that, consciously or not, stylists are beginning to utilise this idea. The exhaust pipes end in a space separated from the rear base by a grid which diffuses the gases and slows them to the required flow rate where they are discharged into the wake. It is conceivable that this styling trend could eventually be used in modern saloon cars.

Detailed styling discussed below concerns the reduction of the interference drag of a car.

6.4 AERODYNAMIC INTERFERENCE DUE TO SMALL SURFACE PROTRUSIONS

Every car has a number of small items on its external surface, such as wing mirrors, door handles, bonnet ornaments, side lights, radio aerials etc. These protruding items interfere with the general airflow around the car.

In practice it is difficult to measure the actual effect of this aerodynamic interference. In the first stages of car styling, when the scale model is tested in a wind tunnel, these small protrusions when proportionally scaled down are disproportionately small in comparison with the viscous boundary layer formed on the model surface (the aerodynamicist would say that the 'Reynolds number is under-critical and that the resulting aerodynamic effect bears no resemblance to the full sized case). In fact what happens is that the pressure distribution does not resemble real conditions and the measured aerodynamic forces are not the true values. On the other hand, it is difficult to measure the interference effect of small protruding items on a full sized car, as these effects are additive, and we can assess only the total sum of the individual effects. We have, therefore, to rely on the wind tunnel investigations carried out on each full-sized item separately.

Results of the full-sized tests have to be used in assessing the aerodynamic interference drag effects on the rest of the car. The cumulative effect of this interference can be quite considerable and has to be calculated by measuring the drag of each item separately then calculating the drag coefficient with respect to the 'reference area' S_m. Since the total drag coefficient of the car is calculated with respect to the maximum cross-sectional area S, the drag coefficient of each small item must then be converted to the full scale car. The total interference drag due to all the small items calculated in this manner is still not exactly the actual total interference drag effect on the car. We have to add to it a secondary interference drag which is due to the effect of placing a small item in the proximity of the large body. It is best to describe such aerodynamic interference between a small item and the main body of the car to which it is attached by considering the simple example of a single axisymmetric body (Fig. 6.8(a)) to which a small item is fixed at a distance '*e*'. Let us consider these two bodies, first separately, and then together, as one configuration. The small item interferes with the airflow pattern around the larger body and produces a faster flow in the immediate vicinity. This change in the airflow

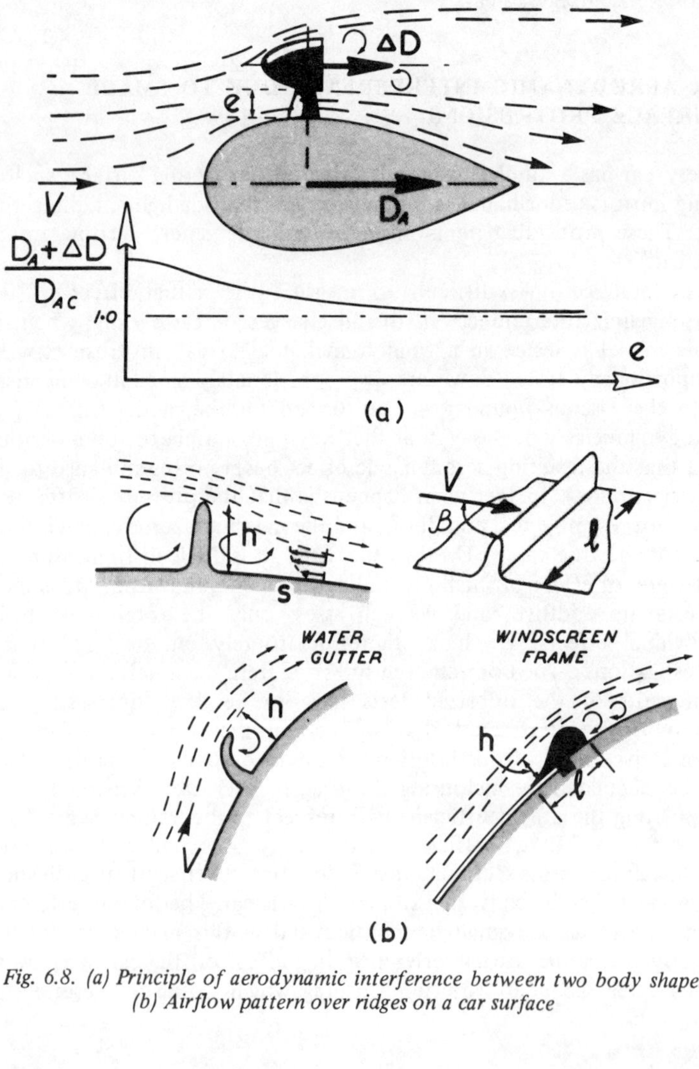

Fig. 6.8. (a) Principle of aerodynamic interference between two body shapes. (b) Airflow pattern over ridges on a car surface

pattern due to the presence of the smaller body causes the additional secondary interference drag which is often quite considerable. We can therefore define 'aerodynamic interference' as the difference between the aerodynamic force of the composite configuration and the sum of the aerodynamic forces due to the component items, but measured separately.

The graph in Fig. 6.8(a) shows the sum of the drag coefficients of the separately measured items divided by the drag coefficient of the composite body

$$\frac{D_A + \Delta D}{D_{AC}}$$

calculated for different distances '*e*'. When '*e*' tends to zero the aerodynamic interference obviously increases. This experiment serves as a good illustration of the aerodynamic interference drag effect of other small items fixed on the car surface.

Side mirror

Consider for example the comparatively simple case of a side mirror. Measuring the drag force D_{Am} of this mirror separately we have

$$D_{Am} = \rho \frac{V_T^2}{2} S_m C_{Dm}$$

where S_m is the maximum frontal area of the mirror, C_{Dm} its drag coefficient and V_T the wind tunnel speed at which measurement is taken. To refer this to the car maximum cross-sectional area S we multiply the drag of the given small item (in this case the side mirror) D_{Am} by the value $1/2\rho V^2 S$, and add to the result the secondary interference, i.e.

$$\Delta C_D = C_{Dm} \frac{V_T^2}{V^2} \frac{S_m}{S} + \Delta\Delta C_D \tag{6.2}$$

where $\Delta\Delta C_D$ is the estimated addition due to the secondary interference. The same method of conversion is applied to all separately measured aerodynamic drag coefficients.

Surface ridge

There are some typical styling features present in practically all types of car. Consider first a ridge, protruding from the surface. Such a ridge can be either a welding seam, or a gutter (as is found in most cars, on the sides of the roof or on the sides of the windscreen). The window frames also can often be considered to constitute a ridge, Fig. 6.8(b). The ridge could be orientated at a different angle to the local airstream and this would greatly affect its interference drag. The base of a ridge is naturally immersed in the viscous boundary layer and only its upper parts are exposed to the moving air. At the same time the presence of a ridge can strongly effect the formation of the boundary layer upstream and downstream of it. Thus the ridge height is of importance. Consider the ridge which is locally normal to the airstream. If the average thickness of the local boundary layer is denoted by δ and the height of the ridge h, then if $h \leqslant \delta$ the effective dynamic pressure (see Equation 2.2) is

$$q_{eff} = \rho \frac{V_e^2}{2} = 0.75 \sqrt[3]{\frac{h}{\delta}}\, q \qquad (6.3)$$

The ridge, as shown in Fig. 6.8(b), causes local flow separation in front and a separated wake downstream of it. When estimating the interference drag increment we refer it to the total frontal area of the ridge S_m, the product of the length and height, and we treat it in the same way as in Equation 6.2. The velocity V_T should be replaced by V_e as given in Equation 6.3. Thus the aerodynamic interference drag increment due to such a ridge is

$$\Delta C_{D\,(ridge)} = \sin^2 \beta \,\frac{0.75}{S} \sqrt[3]{\frac{h^4 x^3}{\delta}} + \Delta\Delta C_D \qquad (6.4)$$

where x is the effective length of the ridge and $\Delta\Delta C_D$ is the estimated secondary interference. Equations 6.3 and 6.4 are valid also for ridges which are formed into open gutters, or have other characteristic shapes such as, for example, window frames.

Surface groove

The engine cover, or bonnet, or any other part of the structure of the car which is hinged on one side and can be opened to give access to internal parts, constitutes another source of interference drag. The outer edges of such pivoted panels usually form a groove in the surface. The aerodynamic effect of such a groove is complicated by the fact that usually the passenger compartment, like the engine compartment to which it provides access, has a different ambient pressure to the flow outside. This is important specifically in the case of the front of the bonnet where the outside airflow is fast and the pressure on the outside surface much lower than inside, where it is near to stagnation. This pressure difference across a panel causes an air leak through the groove. The air leak constitutes not only a momentum loss (drag) from the internal flow point of view but also it interferes with the outside flow thus increasing the interference drag increment. It is very difficult to estimate the effect of this pressure step, and consequent leak, on the interference drag of such grooves.

Consider, therefore, as a first approximation, the interference drag of a groove neglecting the leak effect. Assume also for simplicity that the local airflow is at right angles to the groove. The reference area of the groove is *bl*. The increment of the interference drag coefficient is dependent on the ratio *h*/*b*; experimental results from Fig. 6.9(b) show

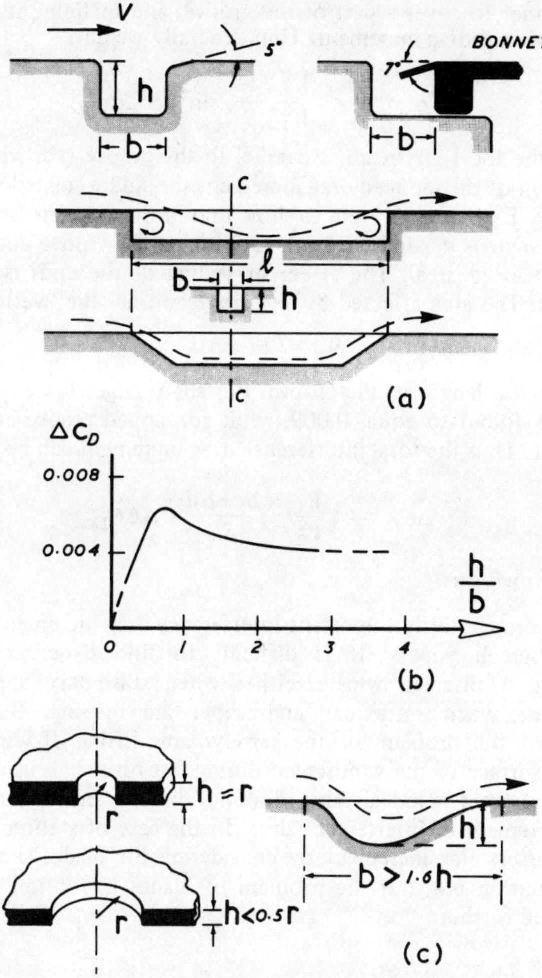

Fig. 6.9. (a) Airflow pattern over grooves in a car surface; (b) effect of groove dimensions on the interference drag; (c) relationship between geometry of holes and surface dents and the interference drag

that at $h/b = 0.6$ this increment is a maximum and amounts to $k = 0.007$, (see Equation 6.5) based on unit length of groove. As the local disturbance caused by the groove mainly affects the viscous boundary layer, and the groove acts as a cavity in which eddies form, the shape of its edges is of aerodynamic importance. Lowering the downstream edge and inclining it at about 5° (Fig. 6.9(a)), decreases the interference drag to about $k = 0.0045$. Extending the external edge of the bonnet to cover a part of the groove, and inclining it inwardly, also reduces this drag increment. Thus generally we have

$$\Delta C_{D\,max} = k \cdot \frac{V_1^2}{V_2^2} \cdot \frac{bl}{S} \cdot \sin^2\beta + \Delta\Delta C_D \qquad (6.5)$$

When the local airstream is parallel to the groove (i.e. when $\beta = 0$) the problem of the induced drag increment should be treated somewhat differently. Experiments seem to show that in this case the interference drag increment is dependent on the styling of the groove ends and the so-called 'wetted area'. The effect of styling of the ends is shown in Fig. 6.9(a). The area affected by the airstream, i.e. the 'wetted area' is

$$S_m = (2h + b)l$$

where l is the length of the groove. For sharp edges the experimental factor was found to equal 0.009, while for sloped groove ends it was only 0.001. Thus the total interference drag increment can be expressed as

$$\Delta C_D = k \cdot \frac{V_1^2}{V_2^2} \cdot \frac{(2h + b)l}{S} + \Delta\Delta C_D \qquad (6.6)$$

Windscreen wipers

Another comparatively important interference drag increment is due to the windscreen wipers. It is difficult to find experimental data concerning windscreen wipers, either when stationary and the car moving, or when both car and wiper are moving. Barth [54] investigated the problem of the aerodynamic lifting of wiper blades from the surface of the windscreen during fast runs. This investigation, although of great practical value, does not directly answer the question of the increment of interference drag. In the case of stationary wipers we may assess this increment by considering the blades as ridges and apply Equation 6.4, but the problem of blades in motion should be investigated further.

Holes and dents in the surface

The surface of a car may have some holes cut in it, or it can have hollow dents such as those connected with door handles. In the case of a hole we may have a pressure difference across the panel which would cause air either to flow in or out; in either case the flow would interfere with the local airstream. It would affect the airstream pattern for a considerable area around the hole and the interference drag increment would be considerable, sometimes larger than the apparent size of the hole.

There is as yet practically no data available on the effect of such an efflux of air on the interference drag. The following considerations are made under the assumption of a simple case of zero pressure difference across the hole area. Consider the area of the hole as the reference area S_m. The interference is affected by the formation of small eddies within the hole, and thus the depth of it is an important factor. When the hole is comparatively 'deep' that is when the effective thickness of the car surface material is of order of the hole radius, then the average interference drag increment is found to be about 0.017. For comparatively larger holes, i.e. when the depth is about half of the radius of the hole, this increment is smaller, i.e. about 0.008. Therefore converting these values to the car reference area S we write

$$\Delta C_D = 0.017 \frac{V_1^2 \pi r^2}{V^2 S} + \Delta\Delta C_D \tag{6.7}$$

where V_1 is the local airstream velocity and $\Delta\Delta C_D$ represent the additional interference effect.

In the case of dents and hollows, Fig. 6.9(c), reference area $S_m = hb$, the experimental value of the interference drag increment is, on average, approximately 0.04. Thus the increment can be expressed as

$$\Delta C_D = 0.04 \frac{V_1^2}{V^2} \frac{hb}{S} + \Delta\Delta C_D \tag{6.8}$$

Composite effect of interference due to small protruding items

When the composite effect of all these styling items are summed up it amounts, for an average saloon car, to about 3% of the total drag coefficient. Available data [9,30,55,56,57] indicates that following these general principles of styling, i.e. avoiding small protruding ridges, setting windows flush in the external surface, smoothing shallow dents, covering all grooves and holes, etc. would reduce the interference drag coefficient by about half, that is by about 1.5%.

This brief survey of the problem of the aerodynamic interference of small features shows that, apart from direct interference, they sometimes have a much greater and often unexpected influence on the aerodynamic characteristics of a given car.

The data in Table 1 has been compiled by looking realistically at different sources of aerodynamic drag and roughly estimating the possible percentage reduction by current styling.

Table 1. Sources of aerodynamic drag

Drag Source	Estimation of the % of total drag	Estimation of possible reduction %	
Form drag	62	base drag	10
		shape	5
Induced drag	10		9
Interference drag	16	underbody	10
		wheels	4.5
		small items	1.5
Internal flow drag	12		6
Total	100		47

This table gives a very approximate indication of the possibilities of drag reduction by styling. Figures quoted are only tentative as details of many of the styling effects are only known superficially and require further extensive experimental investigation.

A school of thought exists among stylists and car engineers that as the final result of styling is dependent on so many different but interrelated influences it is practically impossible to formulate any general principles of styling. Consequently, every car project has to be solved individually by moulding its shape during the preliminary stages of design until more or less satisfactory aerodynamic characteristics are achieved. This is the 'rule of thumb' method which is still occassionally used.

It is hoped nevertheless that this book will succeed in showing that there are solid and fundamental aerodynamic principles which, when followed help in avoiding major mistakes in styling which can result in upset car performance or, worse still, bad handling and stability characteristics.

The designer and stylist usually begins with the required maximum cross-sectional area of the car. Styling the car shape in front and at the rear of the maximum cross-sectional area present the stylist-designer with different and new aerodynamic and design requirements. In solving the problem he can be guided by the ideal shape. Thus far the problem seems to be comparatively straightforward, in practice it is often exceedingly difficult. The suspension mechanism, steering, transmission, the requirements for the internal cooling flow inlets and outlets, drivers visibility, etc. are often complicated and even contradictory. The resulting compilation tends to force designers and stylists increasingly further away from the ideal shape. This requires from the stylist-designer an effort of compromise between conflicting demands in blending together the composite parts of the vehicles shape in the best way possible.

Preliminary estimate of aerodynamic drag

It is useful to state two general principles of styling aimed at reducing the drag, and particularly the interference between different features of the shape.

(1) The airflow around the front should be free to move sideways and not forced to flow over the top and underside. The front planform of the vehicle is better rounded than rectangular.

(2) The rear of the vehicle should be formed so that the cross-sectional area decreases gradually, at an aerodynamically suitable angle. This shaping should avoid sudden changes in inclination of the rearward surfaces to prevent premature separation of airflow. Nevertheless the rear base area should be made as small as possible.

Stylists, when considering a new car project, must follow the designer who imposes the fundamental structural requirements. To check whether the aerodynamic characteristics agree with the expected car performance (power, speed, acceleration) the stylist is often asked, even before the final shape is established, to quote possible aerodynamic coefficients for the car. An estimation of the appropriate aerodynamic drag, or lift, during the preliminary stages of styling is then of decided practical importance since it allows a preliminary assessment of the performance and validates the choice of shape.

Some workers in this field (White [35]) suggest a method based on the statistical rating of given shape characteristics. This is an effort to correlate the particular shape features with resultant drag, or lift coefficients. Dividing shape features into groups such as

(1) The front plan outline.
(2) The front elevation.
(3) The curvature of the windscreen and type of windscreen junction with the body.
(4) The windscreen top shape.
(5) The planform of the canopy top.
(6) The underside panel shape, etc.

it is possible to allocate to each a drag or lift rating value. The stylist is referred to the drag rating table [35] in which the characteristic features of the shape can be compared. The sum of the drag (or lift) rating values, n_d, can then be estimated. An empirical formula for the drag coefficient of such new designs could then be used:

$$C_D = 0.16 + 0.009n_d \tag{6.9}$$

The average margin of error estimated by the authors of this method is approximately 7%.

These preliminary estimates should never replace the actual tests on a scale model in a wind tunnel which are of primary importance. Many styling departments of large car manufacturers study the shape of the car project on a full-sized or scaled-down hard clay model. The measurement of C_D made both on full-sized vehicles and on scale models can be presented in statistical form. Statistics help in this case as it is then possible to make a relative evaluation of the measured data with the average. Statistical values for different types and groups of cars are given in Fig. 6.6(b).

Probably the single event which has had the greatest effect on making motor vehicle manufacturers and users integrate good aerodynamic concepts with the stylists art was the petroleum crisis of the winter 1973/74. In a period of about 14 months the real price of petrol in U.K. increased by 50% and in other countries, due to different levels of taxation on motor fuels, the proportionate increase to the consumer was greater. People realised that oil resources were of finite extent and that fuel prices would inevitably continue to rise as supplies dwindled and new discoveries became more difficult and costly to find and develop. Fuel economy and conservation became matters of both social and economic importance and the transport industry in general responded by recognising that improvements in detailing which made good aerodynamic sense also made good economic sense to the operator, particularly when considered over the life of a vehicle.

Energy savings are particularly relevant to the users of large commercial vehicles and there has been a significant move to increase the permitted maximum size or load capacity of such vehicles. The general principle can be easily illustrated by considering the fuel used in moving say 20 tonnes of goods using (a) four journeys of a 10 tonne g.v.w. lorry with fuel consumption of 15 m.p.g. (18.8 l/100 km) or (b) two journeys of a 16 tonne lorry with fuel consumption of 11 m.p.g. (25.6 l/100 km) and (c) one journey of a 32 tonne lorry with consumption of 8 m.p.g. (35.5 l/100 km). The amount of fuel used is in the ratio of 2.13 : 1.45 : 1 and there are additional economic advantages in the lower standing costs faced by the operator in using fewer vehicles and drivers. This illustration does not include any allowances for empty or part load running, idle time or road structure and environmental damage.

At a road speed of 60 m.p.h. (96 km/hr), approximately half the engine power of a 32 tonne articulated lorry is used to overcome aerodynamic drag, and for each 10% reduction in aerodynamic drag coefficient, the average operator of such a vehicle would save about 500 gallons (2250 litres) of fuel per annum. Various adjustable air deflectors which mount on the cab roof are readily available and experience has shown that they give an average of 6% fuel saving, but careful selection and fitting of a suitable deflector can give 20% fuel savings. Cab deflectors are but one way of reducing the aerodynamic drag of lorries. Details such as radiusing the front edges of the tractor cab, fitting an airdam beneath the front of the cab and local deflector plates to the cab edges and/or the front edges of the load fairing and optimising the gap at the rear of the cab, fitting a vortex stabiliser to the front face of a high trailer, and paying attention to wheel arches and mirrors can all be considered by body builders and operators and help to reduce the amount of fuel consumed by road transport.

There has been an equal impact on the design of saloon and other cars. At an International Symposium held at General Motors Research Laboratories, Michigan, in September 1976, W.-H. Hucho of Volkswagenwerk, and one of the foremost automobile aerodynamicists in the world, stated [92] that 'a drag coefficient . . . of say $C_D = 0.4$ can be realised with shapes of today. However, on the basis of our present knowledge of vehicle aerodynamics, drag coefficients significantly better than this, say $C_D = 0.3$ or below, will require new body shapes'. In the same paper, Hucho also said 'Within existing constraints and the knowledge available today, $C_D = 0.30$ is out of reach for cars designed under usual time and cost limitations'. So great were the economic and commercial pressures over the next few years that at an International Conference in London in June 1982, Hucho said [103] 'Surprisingly there are still cars in production having drag coefficients of $C_D = 0.50$ or even higher. Considering the state of the art, no new car should be launched having a drag coefficient higher than $C_D = 0.40$. This will still be a conservative figure at the time when . . . vehicles . . . will be on the road having drag coefficients of $C_D = 0.30$ or even lower'. Within a very short time of this statement, V-W and other major manufacturers had production cars on the road with a drag coefficient of about 0.30.

Probably the first major public showing of 'ordinary' saloon cars having a drag coefficient of 0.30 or less was at the 1981 Frankfurt International Auto Fair when half a dozen such cars were on view, and others have followed since that date. It is worth examining the styling and engineering details which have contributed to the low drag development of two of these cars – the Ford Probe III [104] which led to the Sierra, and the development of Peugeot from the model 305 to the VERA [97]. Both took as base vehicle an average European medium-large car with a drag coefficient of 0.44. The final versions of both cars were very smooth and aerodynamically clean, with drip rails deleted and joins and gaps between doors and bodywork being as small as practicable ($\Delta C_D = -0.015$). Both have faired head lamps and windscreens which are bonded flush with the bodywork ($\Delta C_D = -0.012$). Front airdams and rear spoiler or boot aerofoil are common ($\Delta C_D = -0.040$) as are flush wheel covers ($\Delta C_D = -0.010$), rear wheelarch fairings and carefully designed front wheel arches ($\Delta C_D = -0.012$). Sealing gaps at the front end and around the bonnet of the Peugeot, a modified grille and ducting to the radiator gave a $\Delta C_D = -0.020$ and the addition of an underpan below the rear seats gave a further $\Delta C_D = -0.010$. Deflectors were fitted around the front door windows and extensions were fitted to the rear pillars and roof end which together gave $\Delta C_D = -0.030$, so the drag coefficient for the Peugeot VERA attained 0.291 with radiator closed. The cooling drag coefficient was 0.017, which represented a reduction of 0.026 from that of the model 305. With the Ford Probe III, greater benefit was gained by attending to underbody detail where a drag reduction of $\Delta C_D = -0.06$ was achieved and the final drag coefficient attained was 0.22.

Whilst some of these detail changes, such as closer tolerances on gaps around doors, sealing of bonnet panels and elimination of drip rails, give rise to problems for the production engineer, it is clear that on balance the improvements gained are economically worthwhile. It appears most likely that there will be further steady progress in design improvements with an associated reduction of drag coefficients of cars in volume production over the next decade, and a consequent long term saving of fuel.

Chapter 7

FUNDAMENTAL ANALYSIS OF FORCES AFFECTING CAR STABILITY

7.1 HISTORIC DEVELOPMENT

In discussing problems related to the effect of aerodynamic forces on road holding, steering and vehicle stability characteristics, we first have to establish a general terminology.

Over 80 years of car development (1890-1975) have not brought uniformity in the definitions and terminology used in this subject and this is precisely what is required if we are to achieve clarity of expression [52,58,59].

It is especially important to define the basic terms 'stability' and 'road holding'. The term 'road holding' is to some extent self explanatory; the concept expresses the capability of a car to be controlled by its steering mechanism along a trajectory selected and maintained by the driver. It therefore describes the ability of a car to follow the path intended by its driver. Thus the term is basically a static concept, related to maintaining a pre-selected trajectory. Later, when discussing other stability problems, the time related response to any change of steering or external forces will be described.

The term 'vehicle stability' is commonly used, but its precise meaning requires analytical explanation. The term itself is borrowed from aeronautics–it is quite natural that vehicular technology should borrow from aeronautical nomenclature, a process well documented in the history of both these technologies. We have to be very aware of the difference in the precise meaning of this term in aeronautics and in vehicle technology. To grasp this difference we have to look into the historical development of the aeroplane and the car. By definition 'stability' in general dynamics means that a body in motion when disturbed by the action of external forces tends to return to its original motion. This definition has a direct application in aeronautics. When considering an aircraft in flight which is disturbed momentarily by say, a short-lived gust of wind, if after an initial deflection from its attitude and speed it returns to its pre-disturbance attitude and speed we say it is 'stable'. When it does not, we say it is unstable. Thus the term 'stability' in aeronautics means the characteristic ability of an aircraft to come back to its original speed and attitude 'without any action being taken by the pilot', or in other words, automatically. This quality in an aircraft is achieved by shaping it in such a way that any temporary variation in the magnitude and points of application of the aerodynamic forces produce moments which turn the aircraft back to its original flying attitude. In aeronautics we call this 'static stability' distinguishing it from so-called 'dynamic stability' which describes the manner, time and mode by which equilibrium is restored in time-related oscillations. Thus 'dynamic stability' in aeronautics describes the oscillatory behaviour of an aeroplane after its motion has been disturbed. In both static and dynamic stability we consider an aeroplane as a rigid body, i.e. its control surfaces rigidly held (or 'frozen') by the pilot. The problems of manouverability are considered separately.

Research in general problems of stability coincided with the appearance of the aeroplane, and as early as 1911 G.H. Bryan published a work entitled 'Stability in Aviation' which already had its predecessors. In this book the author presented a set of mathematical equations which later became the basis of many other dissertations. As an aeroplane moves in an atmospheric environment, in its early stages of development designers had to overcome the unfamiliar problems of 'three-dimensional' motion. The successful development of the aeroplane depended from the beginning in finding a solution to the problem of control and automatic stability. The history of aviation vividly depicts the efforts to solve these problems. This was achieved in the first years of aeronautics by luck and some common sense, and only later by more analytical methods. Between 1930 and the outbreak of the Second World War, methods of research and design in aeronautics were established. The quick growth in size, increased speed, altitude and duration of flight compelled researchers to study still more deeply the problems of stability.

The automobile appeared on the scene in about 1906, i.e. before the first successful flight, and its historic development was basically different. Road vehicles had their predecessors in horse-drawn vehicles and front steerable wheels had been developed before the first petrol-engined car was designed or invented This kind of control is basically very different from aeronautical control as it is not produced by means of aerodynamic forces acting on control surfaces, but by the frictional forces developed between the wheels and the ground. Thus the automobile worked reasonably well from the very beginning and its development did not depend on the solution of control and stability problems.

This initial apparent success of the automobile resulted in a slowing down of the theoretical work in the field of automobile dynamics, and in consequence a gap was created, albeit temporarily, between the state of theoretical knowledge in aeronautics and vehicular stability.

In the case of the car, when the motion is disturbed by external impulsive forces (such as a gust of wind or a change of road camber), the correction of its motion is not automatic but has to be performed by the driver actuating the steering. Automatic stability, does not exist in automobile dynamics in its full sense, thus we have a basic difference in the fundamental concept of stability. In the case of a car the effect of an external disturbance is corrected by the response of the driver who can vary the wheel-road frictional forces restoring the vehicle to its original attitude.

Our knowledge of automobile dynamics has increased steadily since 1906. The problem of aerodynamic forces acting on a car has been considered since the dawn of aviation (i.e. about 1909) but at first only from the point of view of air resistance or drag. Gradual progress in car design has brought developments in the theory of steering and has seen the establishment of general principles in the mechanics of car motion. At first, all theoretical and fundamental research was concerned exclusively with mechanical behaviour and did not include aerodynamic forces. However as the effect of these aerodynamic forces can only be studied with a prior knowledge of mechanical characteristics, it was indeed fortunate that recent developments in this field had prepared the ground so well.

The study of aerodynamics on the stability and road holding ability of vehicles, in practice, began in about 1930. In Germany, Kamm started the interest in front wheel drives. As this design was rather unorthodox at that time, he felt that the whole field of stability and road holding should be investigated to verify this new concept. In surveying this field we find that the researchers had in fact, for the first time included the effects of aerodynamic forces in their investigations. These investigations were carried out by a group of scientific workers who under Kamm later established the well-known 'Research Institute for Vehicular Science and Engines' in Stuttgart.* Kamm had assumed, following aeronautical practice, that a car should have a certain amount of automatic stability, and that the remainder of the correcting forces should be provided by the driver's steering response. This work was, unfortunately interrupted by political upheavals in Germany which resulted in the direction of resources and research to more military matters than the study of car aerodynamics.†

* This Institute is today affiliated to Stuttgart's Technological University and is one of the main European centres of research in car technology.

† Dr. Kamm emigrated to the U.S.A. and became a member of the Stevens Institute of Technology. After the war he returned to Germany and worked in the Batelle Memorial Institute, but his main contribution to car stability research was made in the U.S.A.

The present high level of achievement in the mechanical side of road holding and stability is the result of over 40 years work. Prominent in this field in the U.S.A. are members of the Cornell Aero Laboratory, the Chrysler and Ford Corporations, and other research institutes [60-64]. In this country the main centres of research are the Motor Industry Research Association, near Nuneaton, and some universities [65,66]. Few published research papers include the effects of aerodynamic forces on stability problems and in investigating the work of others in this field, Hales [65], Russell [68] and the Author [67] have tried to add to the calculations the effect of aerodynamic forces on the basic stability and road holding. The present chapter is based mainly on these efforts and presents the subject in a rather general way.

7.2 FORCES AFFECTING A CAR IN STEADY STATE MOTION

The motion of a vehicle takes place on the ground and is affected and controlled by forces created between the wheels and the surface of the ground. In other words the car is 'guided' along a given trajectory by the configuration of the wheels. Any external forces which disturb the forces in equilibrium, change the conditions of motion and consequently the trajectory. The vehicle in itself does not possess the capability of correcting these changes and its motion is not restored automatically. To return to the original path we have to introduce an additional wheel deflection which will introduce additional frictional forces so restoring the trajectory. Thus we can distinguish between steady state motion and transient state motion.

In the present chapter we consider only steady state motion, i.e. when all the forces and moments acting on the car are in equilibrium. In discussing steady state motion we recognise that road holding and steering behaviour cannot be separated from mechanical effects, such as frictional road-wheel interaction, mass-inertia forces and aerodynamic forces. The effects of these two different kinds of forces acting in conjunction have to be considered in our analysis.

Accepting the meaning of the term 'stability' applied to cars with previously mentioned reservations, and considering first steady state motion, we find that the extent of the steering correction depends not so much on the actual forces involved as on their 'rate of change' with the varying car attitude relative to the air stream.

Designers of the first primitive cars (about 1911) mistakenly assumed that wheels always move in their plane of symmetry. This excluded the possibility of side-slip of the wheel on the road and the

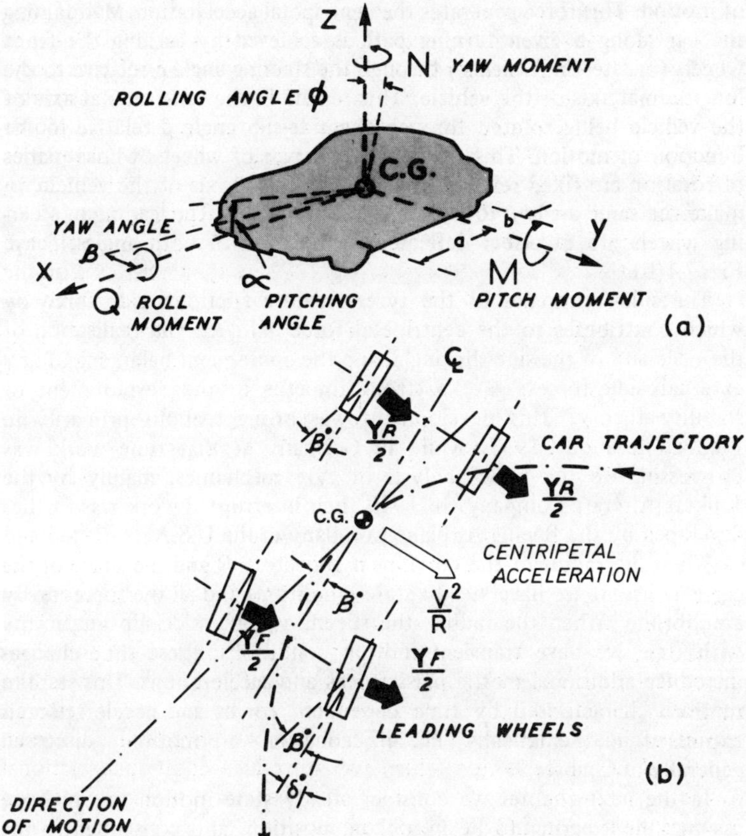

Fig. 7.1. (a) Moments and angular deflections about the centre of gravity. (b) Geometry of turning–sideslip angle and tyre frictional side forces

resulting frictional side force. The concept of the tyre side-slip angle* (Fig. 7.1(b)) related to the car turning trajectory made possible a correct understanding of the role of road-wheel interaction. The side-slip angle β between the direction of motion of a wheel and its plane of symmetry (identical to its plane of rotation) is the inevitable result of turning. For a car to follow a curvilinear trajectory there must be a sideways force acting at right angles to the instantaneous direction

* Evolved in 1925 by Broulhiet, in France, and further developed by work on steering and suspension in 1931 by Becker and others in Germany, and by General Motors in the U.S.A.

of motion. This force generates the centripetal acceleration. Maintaining the car along a given turning path is achieved by turning the front wheels (or 'steering wheels') through the steering angle δ relative to the longitudinal axis of the vehicle. This results in the longitudinal axis of the vehicle being rotated through the side-slip angle β relative to the direction of motion. This causes the rear pair of wheels, whose planes of rotation are fixed relative to the longitudinal axis of the vehicle, to make the same angle β to the direction of motion. The leading or steering wheels are in effect deflected by the sum of both angles, $\delta + \beta$, Fig. 7.1(b).

The side-slip angles of the tyres create a frictional side force Y_T which contributes to the centripetal force MV^2/R. The realisation of the necessity of the side-slip angle, and the consequent balancing of any external side forces, gave a strong impetus to the development of stability theory. This development was brought about primarily in France, in about 1937, while in Germany at that time work was progressing on the basic analysis of tyre mechanics, mainly by the Junkers Aircraft Company. In 1949 their interrupted work was further developed by the Boeing Airplane Company in the U.S.A.

When the radius of the curvilinear trajectory R and the speed of the car is constant we have steady state conditions and all the forces are in equilibrium. When the radius, the speed, and the side-slip angle vary with time we have transient motion conditions; these time changes introduce additional inertia mass forces and accelerations. This state of motion, characterised by time dependent forces and acceleration, is expressed mathematically in an 'equation of motion', discussed generally in Chapter 9.

In the next chapter we consider steady state motion in which we assume the steering to be in a fixed position, and consequently the wheel configuration is assumed constant. We use the term 'stability' to describe the motion behaviour. In the steady state case we mean by stability the 'normal', i.e. controllable behaviour, while the term 'unstable' describes uncontrollable car behaviour. The fundamental criterion of the directional stability of vehicles in steady state motion is the 'critical speed'. At this speed the response of the vehicle due to steering or an applied external force becomes infinitely great, which in practical terms means uncontrollable behaviour.

In this chapter, the analysis of the steady state response to the change of either steering or external forces is expressed as the ratio of two functions (Equation 8.8) of the vehicles characteristic derivatives. The denominator is a function of car speed and the value of the velocity at which the denominator becomes zero is termed the 'critical speed'. As stability in the steady state is related to steering the terms used to describe this type of vehicle response are 'understeer' or 'oversteer'. The criteria involved in this terminology are the trajectory curvature R and the deflection angle δ of the steerings wheels [69,70]. In the case of understeer the effective turning radius R at a given fixed steering angle δ decreases with increase of speed. In the case of oversteer the curvature R increases with increase of speed for a given steering wheel deflection.

An investigation of steady state motion requires an analysis of the characteristic parameters, and their derivatives, which play an important part in car behaviour.

7.3 POSITION OF THE CENTRE OF AERODYNAMIC PRESSURE

The total aerodynamic force created by the relative airstream around the vehicle body may be considered as acting at a point. This point is called the 'centre of pressure', abbreviated C.P. (Fig. 7.2). The aerodynamic force obviously produces no moment about this point but for any other point situated at a distance from the C.P. there will be a net aerodynamic force moment. As all the mass inertia forces acting on a car in motion are concentrated at the centre of gravity it is obviously of interest to know what the aerodynamic moments are which act about the C.G. This requires a knowledge of the relative positions of the C.G. and C.P.

The position of the C.P. is usually found during the preliminary stages of scale model tests in a wind tunnel. The moments about an experimental pivoting point are actually measured for all important model attitudes and the position of the C.P. calculated. In the discussion which follows it will be assumed that the position of the C.P. is known.

7.4 AERODYNAMIC SIDE FORCE Y_A

When the airstream is parallel to the vertical plane of symmetry of the car the resultant aerodynamic force acts in this plane and can be resolved into the usual components of drag and lift. But when the general airstream is inclined to the vertical plane of symmetry the flow pattern is asymmetric and creates an aerodynamic side force. There are usually two main reasons for this asymmetry; first when the car is turning we have the 'side-slip angle', and second when there is an additional atmospheric wind blowing diagonally across the path of the car. In each of these cases the resultant aerodynamic force has to be resolved into three space components, X-force, lift and side force Y_A. The side force acts at the C.P. and is at right angles to the sideways

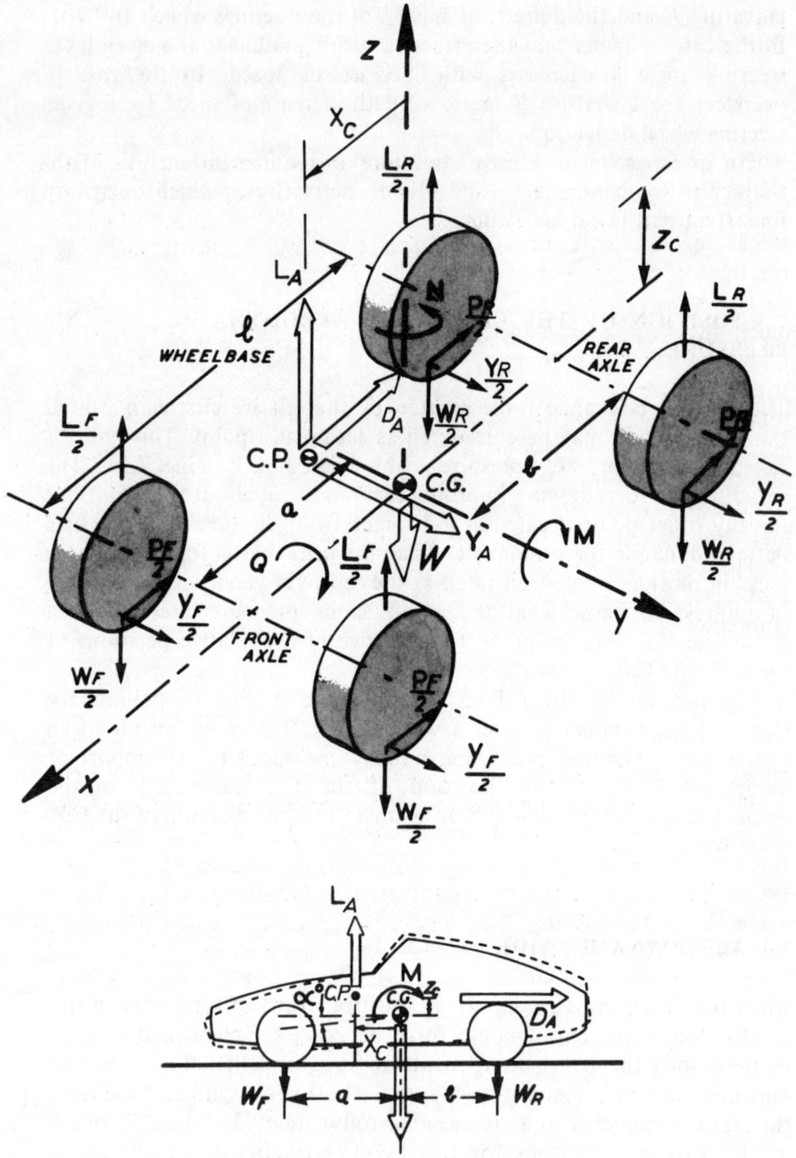

Fig. 7.2. Forces acting at the centre of gravity and centre of pressure, together with road-tyre reaction

plane of symmetry of the car. This force may be expressed in the same form as any other aerodynamical force:

$$Y_A = \rho \frac{V^2}{2} S C_Y \tag{7.1}$$

where C_Y is the non-dimensional side force coefficient. Fig. 7.1(a) shows the axes and angles involved. In all problems related to stability we have to consider the C.G. as the centre of mass motion to which the forces and moments are referred The forces and moments due to the reactions of the wheels are shown in Fig. 7.2.

The aerodynamic side force coefficient of a streamlined car varies much more with the side-slip angle β than do the coefficients of typical saloon cars aerodynamically less well shaped. The value for a streamlined car is $C_{Y\beta} = 2.2$, while for a saloon car $C_{Y\beta} = 1.8$. The total aerodynamic side force Y_A acting on an average saloon car when moving at 70 m.p.h. (114 km/h) in steady side wind conditions can be of order of 250 lbf (1112 N). Racing cars at 180 m.p.h. (290 km/h) could experience a side force of the order of 1000 lbf (4448 N) which may approximate to 25% of their weight.

The detailed analysis of tyre-road reaction, necessary in stability problems, requires resolving the aerodynamic side force into side-component forces acting on the front and rear axles:

$$Y_A = Y_{AF} + Y_{AR} = Y_A \quad \frac{b + x_c}{l} + \frac{a - x_c}{l} \tag{7.2}$$

where x_c is the distance of the C.P. ahead of the C.G., Fig. 7.2.

By considering non-dimensional coefficients instead of the forces the analysis is independent of the actual car size. As the aerodynamic side force is the result of the side-slip angle it is very important, for every typical car shape, to investigate the relation between Y_A and the yaw angle β. This investigation is usually carried out on scale models in wind tunnels. The actual side force is measured for a range of different side (or yaw) angles. It would otherwise be difficult, in practice, to measure the actual aerodynamic side force acting on a full-sized car on the road.

Fig. 7.3(a) shows a typical diagram of the side force coefficient C_Y in terms of the yaw angle β. This diagram shows that a change in β results in a corresponding change in C_Y. This change can be expressed in the form of a derivative

$$\frac{\partial C_Y}{\partial \beta} = C_{Y\beta}$$

which is in fact the slope of the curve $C_Y = f(\beta)$. This rate of change of

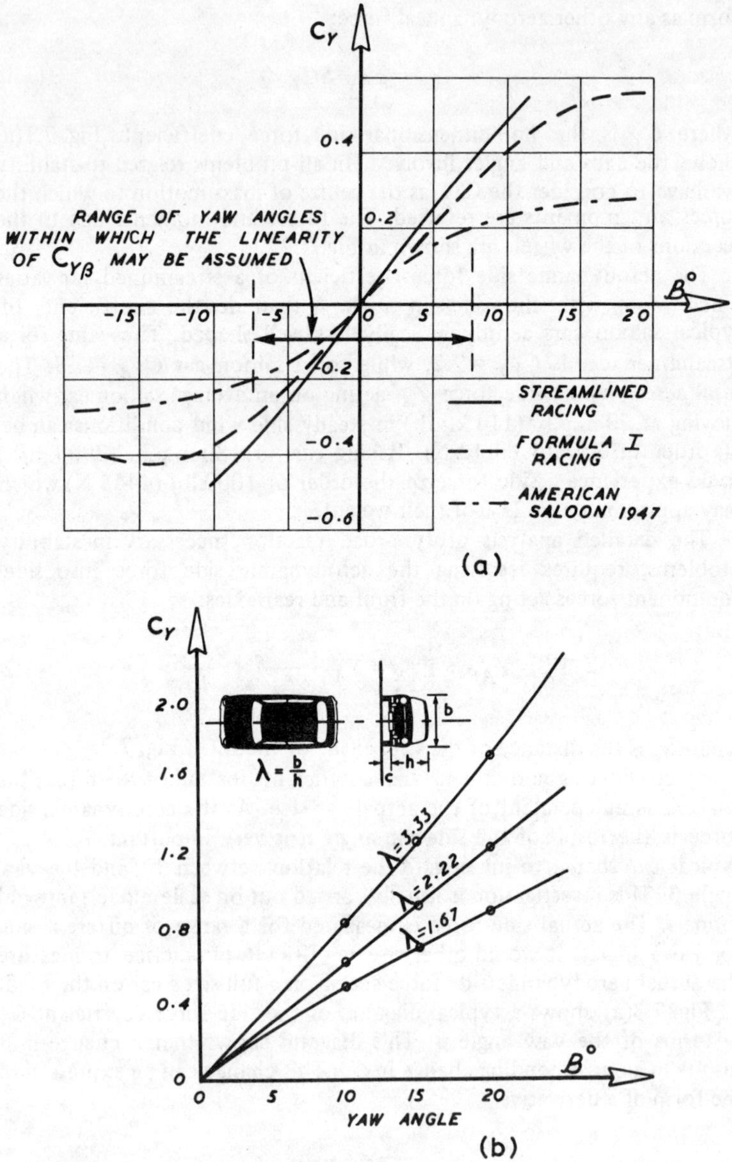

Fig. 7.3. (a) Aerodynamic side force coefficient as a function of the yaw angle; (b) relation between the aerodynamic side force coefficient and the yaw angle for different car sizes

the aerodynamic side force coefficient with the yaw angle is an important characteristic for a given car shape. The curve C_Y versus β is generally not linear and its slope varies with β but, for practical reasons, we consider only a comparatively small range of the angle β close to zero. The curve near $\beta = 0$ is almost linear and in practice we assume that it is linear which allows us to consider the derivative $C_{Y\beta}$ as a constant for a given car.

It is reasonable to assume that the aerodynamic side force is practically independent of the angle of pitch α, which varies only slightly. This same assumption can be made with respect to the variation of C_Y with the roll angle ϕ, thus

$$\frac{\partial C_Y}{\partial \alpha} = \frac{\partial C_Y}{\partial \phi} = 0$$

A preliminary estimation of the aerodynamic effect on stability cannot be made without a knowledge of Y_A versus β, and it is therefore important to estimate the approximate aerodynamic side force characteristics as early as possible during the project stage.

The characteristic dimensions of a vehicle can be expressed as the non-dimensional height to width ratio b/h (Fig. 7.3(b)). The experimental values of C_Y versus β for the three different b/h ratios can help in estimating roughly the order of value of C_Y once the ratio of height to width is determined. This value of C_Y may differ considerably from the one established later due to the secondary interference effects introduced in the final styling of the body.

There are a few practical methods of estimating approximately the side force coefficient. The side force values for shapes with different basic geometrical dimensions has been compiled from accumulated material [71]. An empirical formula has been suggested from this statistical approach which gives the estimated value of the aerodynamic side force coefficient for different yaw angles:

$$C_Y = K\beta \tag{7.3}$$

K is a factor established experimentally for each characteristic group of vehicle shapes where β is measured in radians. This is called the 'Bowman formula' after the author. Statistics show that K is of the order of 2.2 for a typical large American saloon car and about 2.85 for a streamlined racing car [72].

In the U.K. a much better method has been suggested [87] based on the relation between particular characteristics of the car shape to the measured aerodynamic coefficients. We mentioned such a 'rating method' in conjunction with the drag coefficient, Equation 6.9. The side force Y_A can also be expressed by this method which is based on statistical data and considers the shapes in common use–the stylist should be careful not to use it in the case of an unorthodox design. It is intended to be used as a means of estimating Y_A in the early stages of design and cannot be considered as a final and valid value. Any of these empirical methods, therefore, must be validated at a later stage by comprehensive wind tunnel tests on a scale model.

The method is based on a 'rating table' grouping particular features of shapes and allocating to them a 'rating' value. The stylist, having chosen a preferred shape compares its features to those in the table and establishes a composite side force rating factor n_f. By introducing this value into the following formula an approximate value of the side-force coefficient can be found:

$$C_Y = \frac{\text{side area}}{\text{frontal area}}\,\beta\,(0.005 + 0.0019 n_f) \tag{7.4}$$

It is possible to predict the value of C_Y for the average car with an approximately 20% margin of error using this formula. What may be of even greater practical value with this method is that the estimation of the side force coefficient permits the early prediction of the position of the centre of pressure with approximately 5% margin of error.

The aerodynamic side force could also be affected, in theory, by the deflection δ of the leading (or steering) wheels, or by the rotational speed of the car around its vertical axis. This rotational speed is denoted by r rad/s. Thus, in theory, the rate of change of Y_A with either δ or the angular velocity r is

$$\frac{\partial Y_A}{\partial \delta} = Y_{A\delta} \quad \text{or} \quad \frac{\partial Y_A}{\partial r} = Y_{Ar}$$

Little is known about either of these derivatives. It will be assumed that $Y_{Ar} = 0$ and therefore, that this derivative can be neglected.

The Author has tried to measure the value of the derivative $Y_{A\delta}$ using a scale model of a Formula I racing car in a wind tunnel. This type of racing car has its wheels exposed to the airstream and it was assumed that any front wheel deflection would affect the slope of Y_A versus δ. Results of these measurements are given in Fig. 7.4(a). These show that for zero yaw angle the value of $C_{Y\delta} = 1.37\ \text{rad}^{-1}$. The value of this derivative, i.e. the rate of change of the side force with wheel deflection relative to the airstream, was found experimentally by the Author.

The wheels of a typical saloon car rotate within mudguard cavities and are thus effectively shielded from the external airstream. It is to be expected in such a case that $C_{Y\delta} \ll 1$ and is quite insignificant for all practical purposes. Thus in the following calculations, although we retain the derivative $Y_{A\delta}$, it is very small and probably could be neglected.

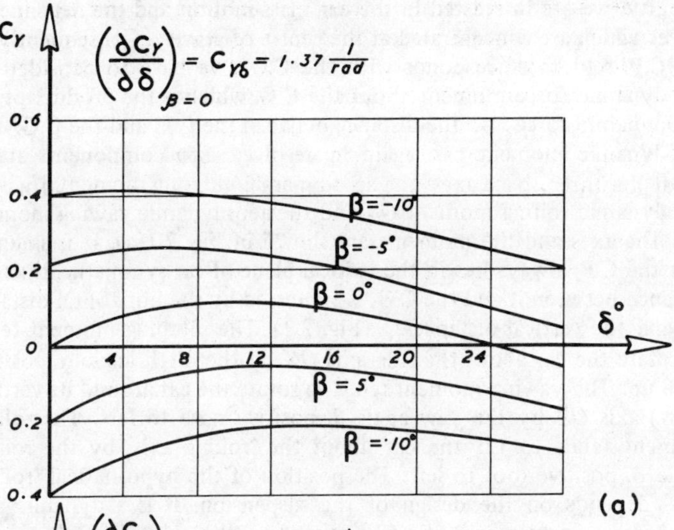

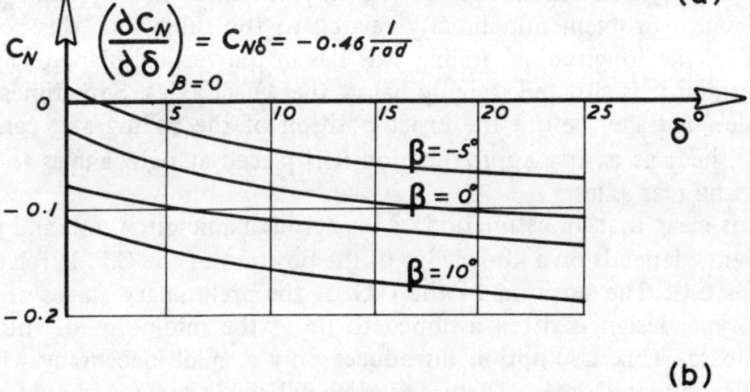

Fig. 7.4. (a) Relation between the aerodynamic side force coefficient and deflection of the steering wheels for different values of yaw angle; (b) relation between the aerodynamic yaw moment coefficient and deflection of the steering wheels for different values of yaw angle

7.5 AERODYNAMIC MOMENTS ABOUT THE CENTRE OF GRAVITY

As the resultant aerodynamic force acts at the centre of pressure it does not produce any moment. At any other point on the vehicle the aerodynamic force produces the so-called 'aerodynamic moment' which is the product of the value of the aerodynamic force and the distance between the C.P. and the given point. In considering problems of stability, we are interested in the car mass-motion and the mass-inertia forces which are concentrated at the centre of gravity. Consequently, as the C.P. seldom corresponds with the C.G., we have to consider the aerodynamic force moment about the C.G. which is the product of the aerodynamic force and the distance between the C.P. and the C.G. This aerodynamic moment can again be resolved into components acting about the three space axes: the aerodynamic pitching moment M_A, the aerodynamic rolling moment Q_A and the aerodynamic yawing moment N_A. The axes and the moments are shown in Fig. 7.1(a). It is assumed that the C.P. always lies in the vertical plane of car symmetry, thus the distance between it and the C.G. is expressed by the horizontal distance x_c and the vertical distance z_c (Fig. 7.2). The pitching moment tends to rotate the car about the side axis OY by the pitch angle α, positive nose up. The yawing moment tends to rotate the car around its vertical (yaw) axis OZ by the yaw angle β, positive nose to left. The rolling moment tends to tilt the car about the 'rolling axis' by the rolling angle ϕ, positive roof to left. The position of the hypothetical 'rolling axis' depends on the design of the suspension. It is a dynamic and mechanical problem not directly related to the subject of this book. Usually, the longitudinal rolling axis lies in the vertical plane of symmetry and it is situated roughly below the wheel axles. Sometimes, at the design stage, before the exact position of the rolling axis can be established, as a first approximation it is placed at right angles to the front and rear axles.

It is clear that an estimation of the aerodynamic pitch, roll and yaw moments depends on a knowledge of the position of the C.P. in relation to the C.G. The position of the C.G. at the preliminary stages of the prototype design is often assumed to be at the mid-point of the car wheelbase. This assumption introduces only a small inaccuracy which can be corrected later. These experimental investigations should give reliable results, showing aerodynamic moments about the C.G. as a function of the relative attitude of the car.

7.6 AERODYNAMIC YAWING MOMENT N_A

This moment is caused by the aerodynamic side force Y_A and we can write

$$N_A = Y_A x_c = qSl\, C_N \tag{7.5}$$

where l is the wheelbase. In all cases of aerodynamic moments the wheelbase is considered to be the *characteristic length.* The yaw moment coefficient is therefore

$$C_N = \frac{N_A}{qSl} \tag{7.6}$$

There is scanty measurement data for full-sized cars. To indicate the actual scale of values involved some measurements on a full-sized saloon car are given in Fig. 7.5(a); the actual value of the aerodynamic yaw moment for fast cars could be about ten times greater. A characteristic curve of the yaw moment coefficient versus the yaw angle β is given in Fig. 7.5(b).

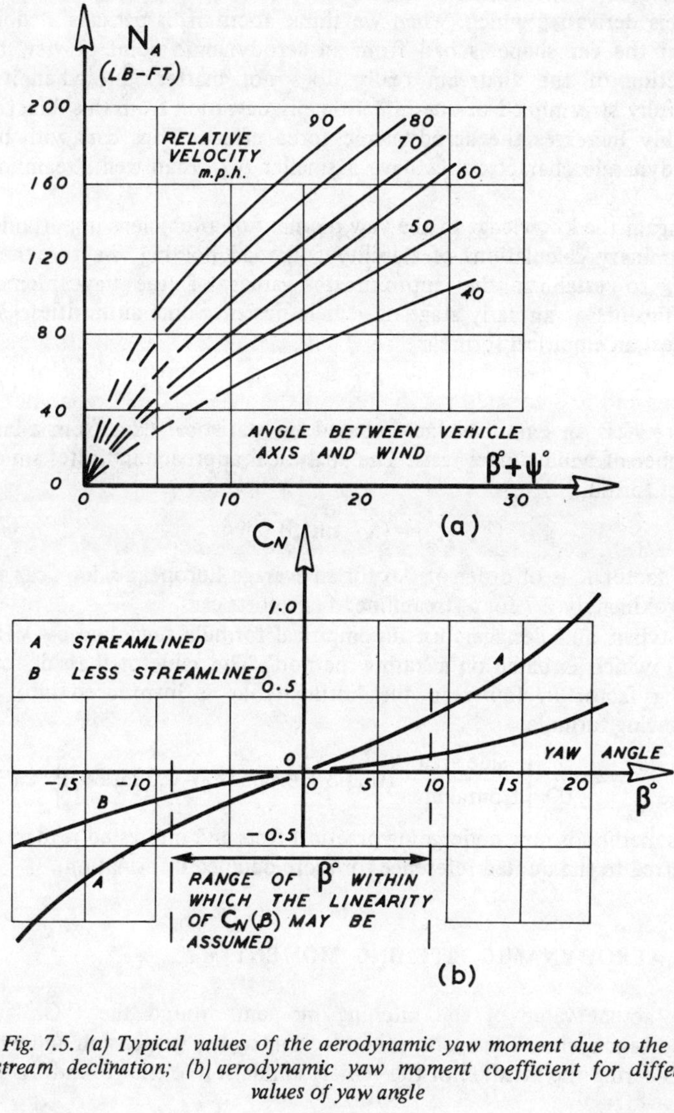

Fig. 7.5. (a) Typical values of the aerodynamic yaw moment due to the side airstream declination; (b) aerodynamic yaw moment coefficient for different values of yaw angle

As in the case of car stability we are more interested in the rate of change of yaw moment than with the value of N_A. The rate of change is the slope of the curve of N_A versus β.

$$\frac{\partial N_A}{\partial \beta} = N_{A\beta} \qquad \text{and} \qquad \frac{\partial C_N}{\partial \beta} = C_{N\beta}$$

Fig. 7.5(b) shows that aerodynamic streamlining results in an increase in this derivative which, when we think about it, is perhaps obvious. When the car shape is bad from an aerodynamic point of view the direction of the airstream really does not matter, but when it is carefully streamlined in one direction any deviation from this direction quickly increases the aerodynamic force effect. Thus cars with bad aerodynamic characteristics have a smaller $N_{A\beta}$ than well streamlined cars.

Again the knowledge of the yaw moment of a vehicle is important in preliminary calculations of stability and road holding. We try, therefore, to establish the approximate values of the yaw moment coefficients at an early stage of vehicle design. Some authorities [72] suggest an empirical formula

$$C_N = K \sin^2 (3\beta) \tag{7.7}$$

where K is an empirical factor based on statistical data from a large number of wind tunnel tests. The analytical approach indicates a more exact formula

$$C_N = (K - \sin^2 \beta) \tan \beta \tag{7.8}$$

The factor K is of order of 0.6 for an average European saloon car and approximately 2.7 for a streamlined fast sports car.

Stylists and designers use an empirical formula suggested by MIRA [87] which is based on a 'rating method'. The value of the side-force rating factor n_f found in the 'rating table' is introduced into the following formula:

$$C_N = \frac{\beta}{100} \left[\frac{\text{side area}}{\text{frontal area}} (0.208 + 0.0655 n_f - 0.00508 n_f^2) \right] \tag{7.9}$$

This method is now undergoing practical tests and interested readers are referred to the quoted reference for more detailed information.

7.7 AERODYNAMIC PITCHING MOMENT M_A

The actual value of the pitching moment around the C.G. is of importance when considering the reaction of the suspension and when considering the attitude of the vehicle in relation to the ground. As was

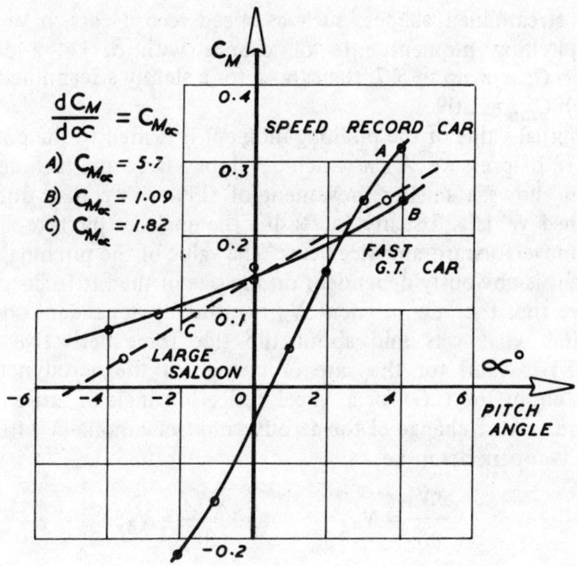

Fig. 7.6. Aerodynamic pitch moment coefficient versus pitch angle

discussed in previous chapters the actual pressure distribution over the external surfaces and in particular on the underside is very sensitive to the pitching angle α, Fig. 7.6. Calculating the pitching moment of the aerodynamic forces around the C.G. (Fig. 7.2) we write

$$M_A = L_A x_c + D_A z_c = qSl\, C_m \tag{7.10}$$

or in a non-dimensional form

$$C_m = \frac{x_c}{l} C_L + \frac{z_c}{l} C_D \tag{7.11}$$

Equation 7.11 allows us to calculate the aerodynamic pitch moment coefficient when the position of the C.G. and C.P. are known. Most investigations concerning the pitch moment are carried out in wind tunnels. In this case, depending on the type of model attachment to the wind tunnel balance, the moment is measured about a pivot point outside the model and the measured value corrected to give the actual moment about the C.G. It is difficult to indicate a typical curve of C_m versus α as this relationship varies a great deal from one car to another. These variations are caused by the ground interference effect; the actual value of M_A depends on the relative position of the C.P. which may wander about due to a change of pitch angle α affecting the flow pattern, especially underneath the vehicle.

Very streamlined shapes, such as speed record cars, have a much higher pitching moment rate of change with α, i.e. a derivative $\partial C_m/\partial\alpha = C_{m\alpha}$ of up to 5.7, than those for a simply streamlined fast car for which $C_{m\alpha} = 1.09$.

The actual value of the pitching moment is related to the position of the centre of pressure. A car which develops a large and turbulent wake behind it shows a smaller movement of C.P. with α than does a well streamlined vehicle. Usually increasing the angle of pitch α causes the C.P. to move forward and vice-versa. The value of the pitching moment coefficient is obviously dependent on the size of the lift force.

We see that the yaw moment N_A is primarily dependent on the side force, and what was said about the side force derivative Y_A (see page 168) is valid for the rate of change of the aerodynamic yaw moment about the C.G. for a wheel deflection angle δ. Little is known about the rate of change of the aerodynamic yaw moment with respect to these two parameter, i.e.

$$\frac{\partial N_A}{\partial \delta} = N_{A\delta} \qquad \text{and} \quad \frac{\partial N_A}{\partial r} = N_{Ar}$$

It is highly probable that as $Y_{Ar} = 0$, $N_{Ar} = 0$ and consequently in the calculations which follow this derivative is neglected.

The case of wheels exposed to the airstream was experimentally investigated (Author's unpublished work) and the yaw moment around the C.G. due to the wheel deflection angle was measured. Results show that when the front wheels are not shielded, e.g. as in a Formula I racing car, it would probably be of the order of $C_{N\delta} = -0.46 \text{ rad}^{-1}$, see Fig. 7.4(b). In the following formulae we retain this derivative, but we should keep in mind that for a typical saloon car which has wheels shielded from the outside airstream, the value of this derivative would be negligible, if not zero.

7.8 AERODYNAMIC ROLLING MOMENT Q_A

The aerodynamic rolling moment is created by the aerodynamic side force Y_A. Its effect is important when evaluating stability and road holding characteristics as this force occurs when the car is turning in still air or when there is a side wind. The rolling angle due to the aerodynamic rolling moment can be additive to that due to the centripetal acceleration effects generated by the suspension on the sprung mass of the vehicle when, as is usual, the centre of mass is above the roll axis. The combined effects on the turning attitude are shown in Fig. 7.7(a).

We have, therefore, to consider the rolling angle ϕ between the suspended mass of the car and its unsprung mass. The total mass of the car is composed of the unsprung mass, comprising wheels and axles

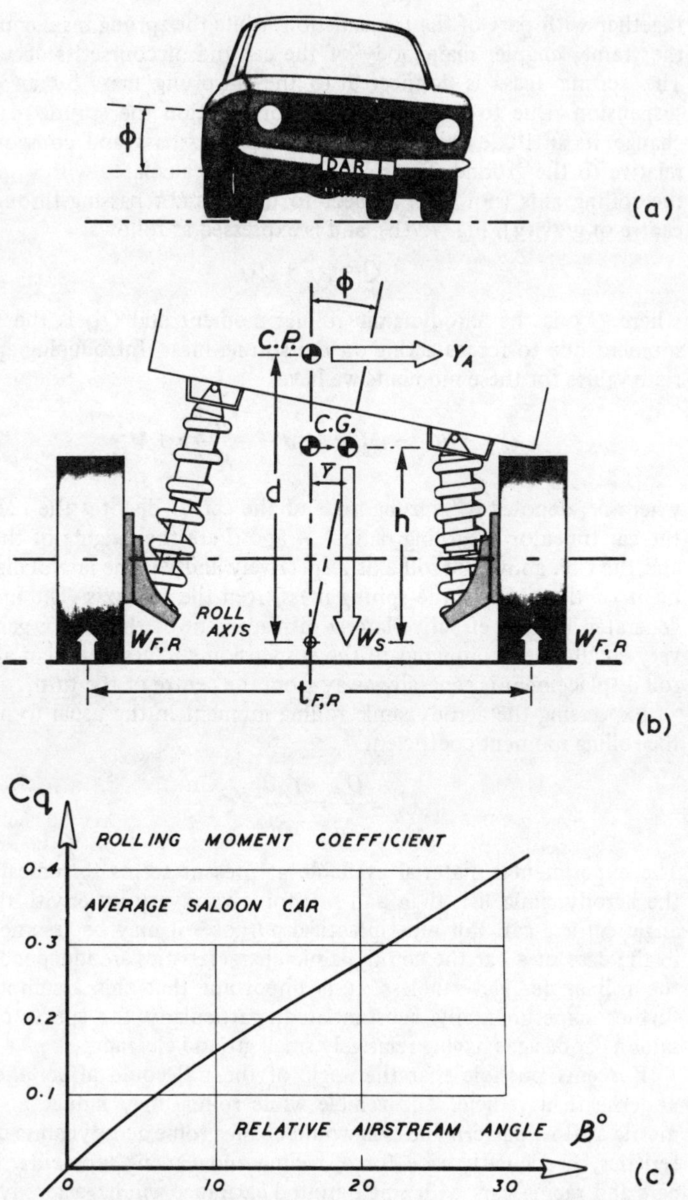

Fig. 7.7. (a) Rolling angle deflection during turning. (b) Schematic representation of an elastic suspension and rolling axis. (c) The rolling moment coefficient as a function of yaw angle

together with part of the transmission, while the sprung mass consist of the frame, engine, main body of the car and of course its occupants. The sprung mass is connected to the unsprung mass by an elastic suspension. Due to this method of construction the sprung mass can change its attitude relative to the unsprung mass and consequently relative to the ground. The rolling moment is calculated with respect to the rolling axis (not with respect to the axis OX passing through the centre of gravity), Fig. 7.7(b), and is expressed as follows:

$$Q = Q_A + Q_M$$

where Q_A is the aerodynamic rolling moment and Q_M is the rolling moment due to forces acting on the sprung mass. Introducing appropriate values for these moments we have

$$Q = qSlC_q + Q_M = Y_A d - \frac{M_s V^2}{R} h + M_s g\bar{y} \tag{7.12}$$

where M_s denotes the sprung mass of the car, R denotes the radius of the car trajectory (turning radius), h and d are the heights of the C.G. and the C.P. above the roll axis respectively and $\bar{y}$ is the lateral displacement of the C.G. of the sprung mass from the roll axis. The moment generated by this effective lateral displacement of the C.G. is generally very small when compared to the suspension stiffness moment and the roll displacement is generally away from the centre of the turn.

Expressing the aerodynamic rolling moment in the usual form gives the rolling moment coefficient

$$C_q = \frac{Q_A}{\rho Sl} = \frac{Y_A d}{\rho Sl} = C_Y \frac{d}{l} \tag{7.13}$$

The experimental material available at present seems to indicate that the aerodynamic lift, drag and side-force vary very little with the roll angle ϕ for a car. For most practical purposes it may be assumed that in the case of a car the aerodynamic characteristics are independent of the roll angle. Nevertheless, it is important that this assumption is further experimentally substantiated, particularly for sports cars or saloon car designs having relatively small ground clearance.

It seems possible that the angle of the roll could affect the flow underneath a vehicle. The vehicle while rolling may induce an asymmetric airflow pattern and this would change some aerodynamic characteristics, particularly side force, rolling and yawing moments. Sports cars and racing cars with small ground clearance will have aerodynamic characteristics which are sensitive to the angle of roll.

It may be assumed, however, that in the present state of knowledge, for all practical purposes for ordinary cars, the roll angle derivatives are zero:

$$\frac{\partial C_L}{\partial \phi} = \frac{\partial C_D}{\partial \phi} = \frac{\partial C_Y}{\partial \phi} = \frac{\partial C_q}{\partial \phi} = 0$$

The aerodynamic rolling moment is directly related to the side force which is dependent on the yaw angle, and thus the rolling moment coefficient is dependent on the yaw angle. This relation is shown in Fig. 7.7(c) which shows the curve of C_q versus β for an average saloon car. The slope of this curve is the derivative

$$\frac{\partial C_q}{\partial \beta} = C_{q\beta} = 1.05$$

There is at present little experimental data available and no practical method of estimating the aerodynamic rolling moment coefficient at the early stages of car project development.

A typical attitude of a car when turning and deflected through a roll angle ϕ about the rolling axis is shown in Fig. 7.7(a). If the roll axis could be made to pass through the centre of gravity of the sprung mass, there would be no roll displacement due to inertial effects. The aerodynamic rolling moment on most cars is quite small. For many racing cars with small clearance beneath the underbody surfaces and the ground, many of the aerodynamic forces and moments are very sensitive to changes in ground clearance. Hence the corresponding roll angle derivatives are not zero and it is necessary to severely limit roll angle magnitudes. This is achieved by increasing the torsional stiffness of the suspension and in some cases relying entirely on tyre stiffness.

In its Advanced Passenger Train, British Rail has designed a suspension mechanism which places the effective roll axis above the centre of gravity of the sprung body of each coach. In this way the moment of the centripetal force acts so as to roll the coach so that its centre of gravity is displaced laterally towards the turning centre, and the coach overturning moment at track level in a curve is reduced. In this way high speed trains can be run safely round curves on track which is correctly banked for much lower speed trains.

Chapter 8

EFFECT OF AERODYNAMIC FORCES ON STEADY STATE STABILITY

In considering the conditions in which a car moves on the road we may, for simplicity, assume that the road is level and that there is no wind. The relative velocity of the air and the car is then the same as the relative velocity of the car and the ground. When the speed and the trajectory of the car is maintained constant under these conditions the motion is described as 'steady state'.

We shall begin by considering steady state car motion which is comparatively well documented in the published literature [60-64,73]. In all of the works referred to mainly frictional and inertia forces are considered, and the aerodynamic forces neglected. Only recently has some work been carried out which includes aerodynamic effects [65,67,68,74].

This chapter, following well established lines, is therefore based on the classic considerations of mechanical forces and moments, but with the effect of external aerodynamic forces added. It thus shows their importance in the overall picture of steady state motion.

When we consider a road vehicle in motion we associate with it a level surface which restricts its motion to a plane. The motion is then restricted to two dimensions, neglecting possible motion in the vertical direction. Car dynamics is complicated by the fact that the chassis and wheels, in practice, move in a plane parallel to the ground while the body is elastically sprung and can pitch and roll relative to the chassis. We therefore have to consider the motion of both the sprung and unsprung masses, mechanically and aerodynamically.

The conditions of equilibrium of the forces and moments in steady state motion are expressed in terms of the characteristic derivatives and deflection angles. By assuming small deviations from the condition of equilibrium we can then investigate the controllability of the motion. If the resulting radius of turning, or the yaw angle involved, becomes uncontrollable we say that the car is 'unstable', or that the condition of motion is 'critical'.

There are two possible approaches to the problem of steady state stability (remembering the introductory remarks concerning terminology we use the term 'steady state stability' with the reservations expressed in Section 7.1). The first approach is to write down all the possible conditions of equilibrium of forces and moments, including the effects of the pitch, yaw, and roll deflections. It is a method of analysing the most complicated case by simplifying it gradually, neglecting the less important factors. The second approach is to analyse the most simple case first and then to introduce gradually the more complex factors. This second approach is accepted by most investigators and is the one we follow here.

8.1 EQUATIONS OF EQUILIBRIUM

By neglecting the rolling deflection of the sprung mass and considering only the motion of the vehicle in a plane parallel to the ground, in the first instance, we are restricted to:

(1) Longitudinal motion.
(2) Sideslip motion (lateral).
(3) Yaw angular motion (rotation about the vertical axis).

When the rolling of the sprung mass is neglected we can neglect also the corresponding weight distribution changes. Consequently the port and starboard wheels behave in the same way and can be considered as one. The whole vehicle can be practically represented by 'collapsing' the four wheels into a 'tandem' configuration, Fig. 8.1(a). To simplify matters further we also neglect the effect of the angular pitch deflection which would change the weight distribution between the front and rear axles.

Analysing this simplified motion along a generally curvilinear trajectory it will be seen that to generate the centripetal acceleration we must have frictional wheel-road reactions which are only present when the wheels have a 'side-slip' angle. Thus the whole axis of car symmetry has to be inclined to the side (right or left depending which way we turn) by the yaw angle β. This yaw angle (or, as it is sometimes called 'side-slip angle'), is the main feature of the turning motion. The corresponding side-slip angles of the front axle wheels is ζ_F and the rear axle wheels ζ_R. From the geometry we may conclude that

$$\zeta_R \cong \beta \quad \text{and} \quad \zeta_F \cong \beta + \delta.$$

These side-slip angles give rise to the side-slip frictional force Y_T which contributes to the centripetal force on the turning vehicle. This yaw angle causes the resultant aerodynamic force to deviate from the car plane of symmetry and consequently it produces an aerodynamic side-force component Y_A. In conditions of steady wind there would obviously be an additional reason for the presence of an aerodynamic side-force if the atmospheric wind is inclined to the car trajectory.

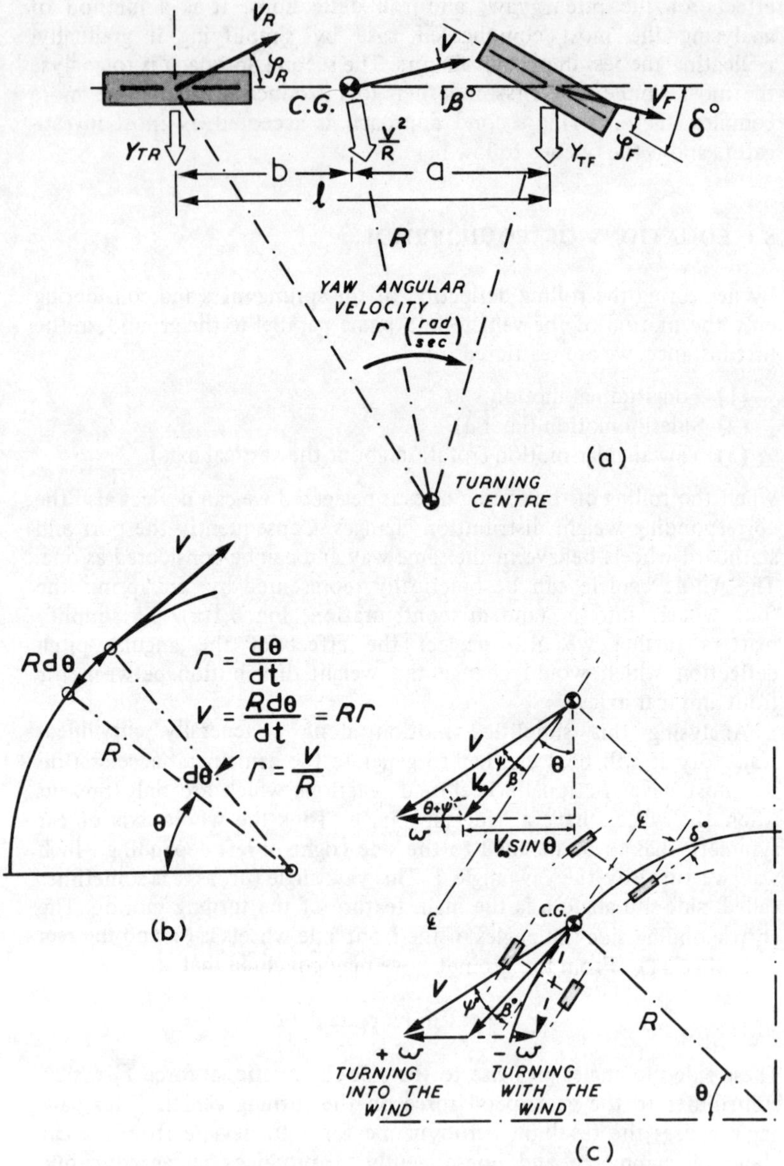

Fig. 8.1. (a) Wheel side-slip attitude when turning neglecting the rolling moment. (b) Relation between tangential and angular velocity during steady turning. (c) Relative direction of the airstream due to car motion combined with wind

Assuming steady state motion we can write for conditions of equilibrium:

forces in the direction of motion (OX)

$$M\frac{dV}{dt} = 0 = T - D_M - D_A \quad (8.1)$$

forces in the side-slip direction (OY)

$$\frac{MV^2}{R} = Y_T + Y_A \quad (8.2)$$

moments around the vertical car axis (OZ)

$$I_z\frac{dr}{dt} = 0 = N_A + N_T \quad (8.3)$$

In these equations, because of the assumption of steady motion, the acceleration $dV/dt = 0$ and the angular acceleration $dr/dt = 0$; I_z represents the 'moment of mass inertia' of the car about the vertical axis OZ.

Consider first Equation 8.1 which states that the force which propels the car with uniform motion in a straight line is equal to the sum of the mechanical drag force D_M plus the aerodynamic drag D_A. This equation expresses only the equilibrium of the drag and thrust forces which are both known initially. It does not help in solving stability problems and this equation is therefore omitted in further discussions.

Equation 8.2 expresses the equilibrium of the 'side-slip' forces, i.e. that the inwards centripetal force MV^2/R is equal to the sum of the tyre side-slip frictional force Y_T and the aerodynamic side-force Y_A. These forces are best expressed in terms of the relevant characteristic values, i.e. the side-slip angle β, the yaw angular velocity r, and the front wheel steering deflection angle δ (we decided earlier to neglect the rolling angle which would otherwise change the side force). As before all the respective characteristic derivatives are assumed constant because they represent linear variations of particular values. Thus the actual aerodynamic side-force may be expressed as the sum of the products of the deflections and derivatives, i.e.

$$Y_A = \beta Y_{A\beta} + \delta Y_{A\delta} + rY_{Ar}.$$

The last item can be neglected since, as we stated before, $Y_{Ar} = 0$. The value of the tyre side-slip force Y_T* is expressed in the same way, and Equation 8.2 can be written as follows:

$$\frac{MV^2}{R} - \beta(Y_{T\beta} + Y_{A\beta}) - r(Y_{Tr}) - \delta(Y_{T\delta} + Y_{A\delta}) = 0 \quad (8.4)$$

* As this book is concerned primarily with aerodynamics, the frictional side forces between tyres and the road are treated only as far as is necessary in explaining car motion. We may express these forces in a similar way to the aerodynamic side forces, i.e. $Y_T = Y_{T\beta}\beta + Y_{Tr}r + Y_{T\delta}\delta$.

Expressing the sum of the moments about the vertical axis passing through the centre of gravity, and equating it to zero:

$$\beta(N_{T\beta} + N_{A\beta}) + r\,N_{Tr} + \delta\,(N_{T\delta} + N_{A\delta}) = 0 \tag{8.5}$$

The value of the angular yaw velocity r may be more conveniently expressed as $r = V/R$ (see Fig. 8.1(b)).

In the case of an atmospheric wind the relative angle between the resultant airstream and the car would be the sum of the side-slip angle β and the additional 'wind angle' ψ. The wind angle depends on the direction of the wind and the tangent to the trajectory curve. Assuming that the wind is steady, we can express the angle ψ in terms of the reference angle between the direction of the wind and the given radius of turning, Fig. 8.1(c). Thus we can investigate the effect of the wind on turning, referred to the position of the car on the road, from some arbitrary selected initial position. This will be found useful later on but at present we shall assume that there is no wind effect.

Using Equations 8.4 and 8.5 we can investigate steady state stability for the following two cases:

(1) A straight-line trajectory–the motion affected by either a slight steering impulse or by an external impulsive aerodynamic force, both requiring correction.

(2) Turning–affected by a steady wind during steady state motion.

In the first case we assume for simplicity that the trajectory reference angle is constant, $\theta = 90°$, and that the aerodynamic force due to the side wind is small. We can then use direct the two equations of equilibrium in an investigation of the vehicle response when travelling in a straight line normal to a weak wind velocity w.

When the deflection of the front steering wheels is zero, i.e. $\delta = 0$ we can write Equations 8.4 and 8.5 as follows:

$$\beta(Y_{T\beta} + Y_{A\beta}) + \frac{w}{V} Y_{A\beta} - \frac{V}{R}(MV - Y_{Tr}) = 0 \tag{8.6}$$

$$\beta(N_{T\beta} + N_{A\beta}) + \frac{w}{V} N_{A\beta} + \frac{V}{R} N_{Tr} = 0 \tag{8.7}$$

The vehicle behaviour is best characterised by the curvature $(1/R)$ of its trajectory. Solving Equations 8.6 and 8.7 as two simultaneous equations with respect to $1/R$ gives

$$\frac{1}{R} = -\frac{w}{V^2}\left[\frac{N_{A\beta}(Y_{T\beta} + Y_{A\beta}) - Y_{A\beta}(N_{T\beta} + N_{A\beta})}{N_{Tr}(Y_{T\beta} + Y_{A\beta}) - (Y_{Tr} - MV)(N_{T\beta} + N_{A\beta})}\right] \tag{8.8}$$

Equation 8.8 shows the effect of the wind component w on the curvature of the trajectory, $1/R$.

Solving Equations 8.6 and 8.7 with respect to the side-slip angle β in terms of the same wind component gives

$$\beta = -\frac{w}{V}\left[\frac{N_{A\beta}(MV - Y_{Tr}) - Y_{A\beta}N_{Tr}}{N_{Tr}(Y_{T\beta} + Y_{A\beta}) - (Y_{Tr} - MV)(N_{T\beta} + N_{A\beta})}\right] \tag{8.9}$$

It should be noted that in both Equations 8.8 and 8.9 the denominator is identical.

The practical use of these equations is that they express the relationship between the radius of the trajectory or the side-slip angle, in terms of the side wind component w for different (steady) values of the car speed V. We are able to study the effect of car speed by substituting the appropriate derivatives which are constant characteristics for a given car. A numerical investigation based on the average values of tyre and aerodynamic derivatives [68] is shown in Fig. 8.2(a) and (b). We notice that when the common denominator tends to zero, the values of both $1/R$ and side-slip angle β tend to become infinitely great. This means that the radius R becomes zero and the angle β infinitely large, i.e. the car in motion under the influence of any small side wind would become uncontrollable and the car would be unstable. The speed of the car at which this would happen, V_{crit} may be calculated by equating the denominator to zero*

$$V_{crit} = \frac{Y_{Tr}(N_{T\beta} + N_{A\beta}) - N_{Tr}(Y_{T\beta} + Y_{A\beta})}{M(N_{T\beta} + N_{A\beta})} \tag{8.10}$$

A typical case of such behaviour, when the car speed increases towards this critical value, is given in Fig. 8.2(a) and (b). It shows the discontinuity of both the radius of the car trajectory R and the yaw angle β. In practice, designs are checked to ensure that V_{crit} is much greater than the maximum speed the car may achieve. Thus the 'safety margin' of a road vehicle means its speed spectrum which excludes speeds within a sufficiently wide range around the V_{crit}. These stability and road holding investigations were, in the past carried out by considering only frictional and mechanical car characteristics, totally neglecting aerodynamic considerations. Here the aerodynamic derivatives $Y_{A\beta}$ and $N_{A\beta}$ are included and as their values for an average saloon car are of consequence, that is of the same order of value as the tyre friction derivatives, we may conclude that the inclusion of aerodynamic characteristics are very important from the stability point of view.

This problem may be approached from yet a different angle. We assumed above a small side wind velocity and investigated the behaviour

* Similar expressions for this critical speed have been developed by other authors, but excluding the aerodynamic terms and derivatives [62,63].

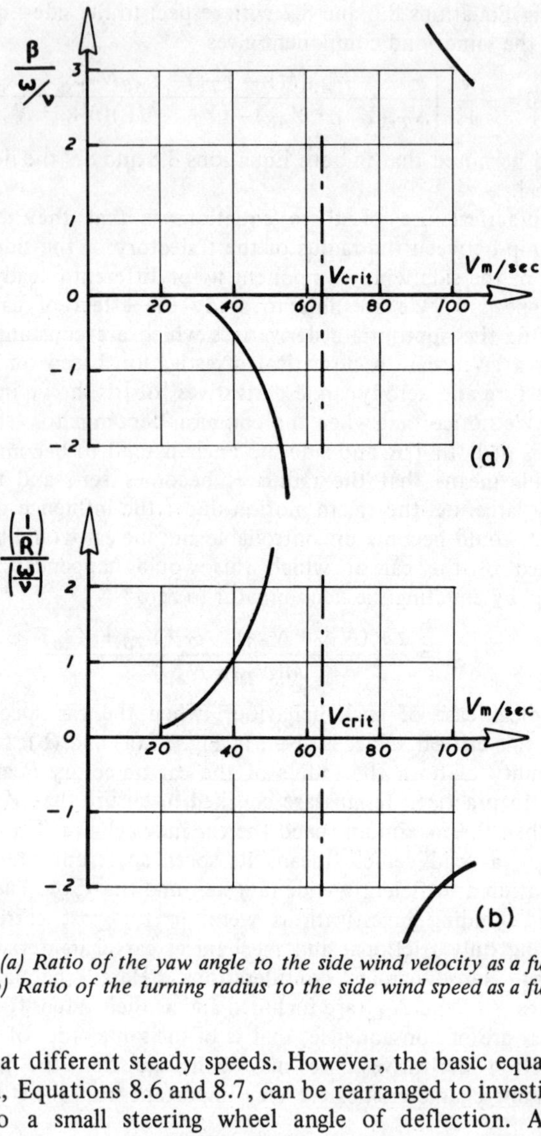

Fig. 8.2. (a) Ratio of the yaw angle to the side wind velocity as a function of car speed. (b) Ratio of the turning radius to the side wind speed as a function of car speed

of the car at different steady speeds. However, the basic equations of equilibrium, Equations 8.6 and 8.7, can be rearranged to investigate the response to a small steering wheel angle of deflection. Assuming therefore that the car travels at a steady speed, that the front steering wheels are turned through an angle δ, and that the atmospheric wind $w = 0$, we can investigate the dependence of the yaw angle on the angle of deflection δ of the front steering wheels. Eliminating the radius R of the trajectory from Equations 8.6 and 8.7 we obtain the yaw angle

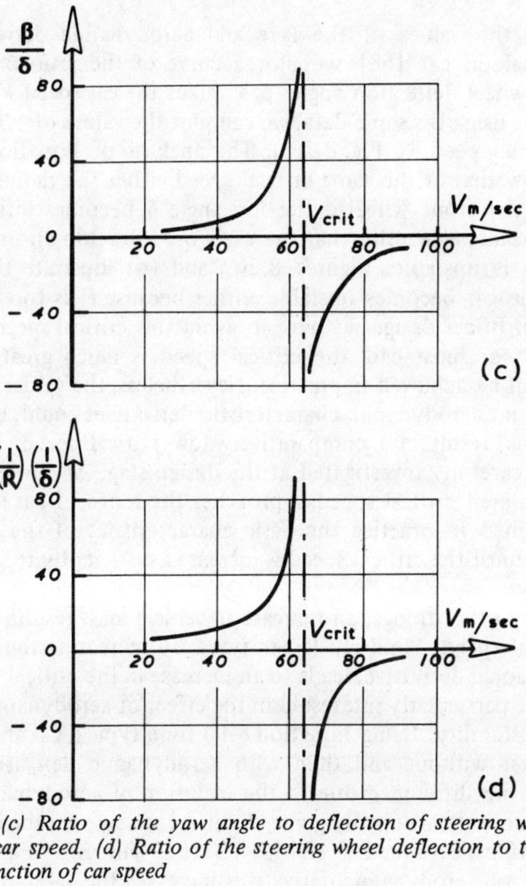

Fig. 8.2. (c) Ratio of the yaw angle to deflection of steering wheels as a function of car speed. (d) Ratio of the steering wheel deflection to the turning radius as a function of car speed

$$\beta = -\delta \left[\frac{N_{Tr}(Y_{T\delta} + Y_{A\delta}) + (MV - Y_{Tr})(N_{T\delta} + N_{A\delta})}{N_{Tr}(Y_{T\beta} + Y_{A\beta}) - (Y_{Tr} - MV)(N_{T\beta} + N_{A\beta})} \right] \quad (8.11)$$

Eliminating in turn the side-slip angle β from both equations we calculate the radius in terms of the deflection angle δ of the front steering wheels. This is, in practice, the most interesting relationship.

$$\frac{1}{R} = -\frac{\delta}{V} \left[\frac{(Y_{T\beta} + Y_{A\beta})(N_{T\delta} + N_{A\delta}) - (Y_{T\delta} + Y_{A\delta})(N_{T\beta} + N_{A\beta})}{N_{Tr}(Y_{T\beta} + Y_{A\beta}) - (Y_{Tr} - MV)(N_{T\beta} + N_{A\beta})} \right] \quad (8.12)$$

Note that as the steering angle δ is involved we include here the derivatives $Y_{A\delta}$ and $N_{A\delta}$ although their values are probably small (at least for saloon cars with shielded wheels).

Using the values of the tyre and aerodynamic derivatives for a typical saloon car [68] we plot a curve of the ratio of side-slip to steering wheel deflection angles β/δ versus the car speed V, Fig. 8.2(c). Similarly, using the same data, we can plot the values of $1/R.\ 1/\delta$ versus the vehicle speed V, Fig. 8.2(d). The analysis of Equations 8.11 and 8.12 show that at the same critical speed either the radius R becomes zero or the front wheel deflection angle δ becomes infinitely great, which means in practice that the car would develop an uncontrollable spin and is unstable. Figures 8.2(c) and (d) illustrate the speeds at which the car becomes unstable, either because R is too small or δ is too large. In car design we have to avoid this critical speed at all costs. However for most cars the critical speed is much greater than that which can be achieved in practice. Nevertheless, the values of both tyre friction and aerodynamic characteristic derivatives could, under certain conditions, result in a comparatively low critical speed. These values must be carefully investigated at the design stage of any car project. If the calculated critical speed approaches the range of car speeds which are obtained in practice the basic characteristics of the car must be changed until the critical speed is increased sufficiently to give a margin for safety.

Among other things, an increase in vehicle mass results in a decrease of critical speed. Similarly larger tyres, by increasing the value of the tyre frictional derivative, leads to an increase in the critical speed.

We are particularly interested in the effect of aerodynamic characteristics on stability. Using Equation 8.10 for a typical car and calculating V_{crit}, first without and then with aerodynamic derivatives, we can estimate the difference due to the inclusion of aerodynamic forces in stability considerations. By giving V different values in the denominator of Equations 8.11 and 8.12 this effect is shown in Fig. 8.3(a). The inclusion of aerodynamic derivatives increases the value of the critical speed and shows previous estimates to be unnecessarily pessimistic. Nevertheless only an exact analysis which includes frictional, inertia and aerodynamic forces and characteristics can assure a sufficient margin of safety for a given type of car.

8.2 THE STATIC MARGIN

It is obvious from basic mechanical principles that when the resultant force of all four wheel frictional side forces acts in the plane passing through the car centre of gravity, and we neglect the rolling moment, the yawing moment due to this force would be zero. Any increase of the velocity, and consequently of the size of the forces, could not result in an angular yawing motion but could only introduce side-slip motion.

This can happen when the sum of the frictional tyre forces cannot generate the required centripetal acceleration. In the usual case when the resultant of all the tyre side forces act along the line which is at a distance x_T from the centre of gravity, Fig. 8.3(b), there will also be a yawing moment N_T. In such a case the distance x_T is a characteristic related to the steady state stability of the vehicle. This characteristic

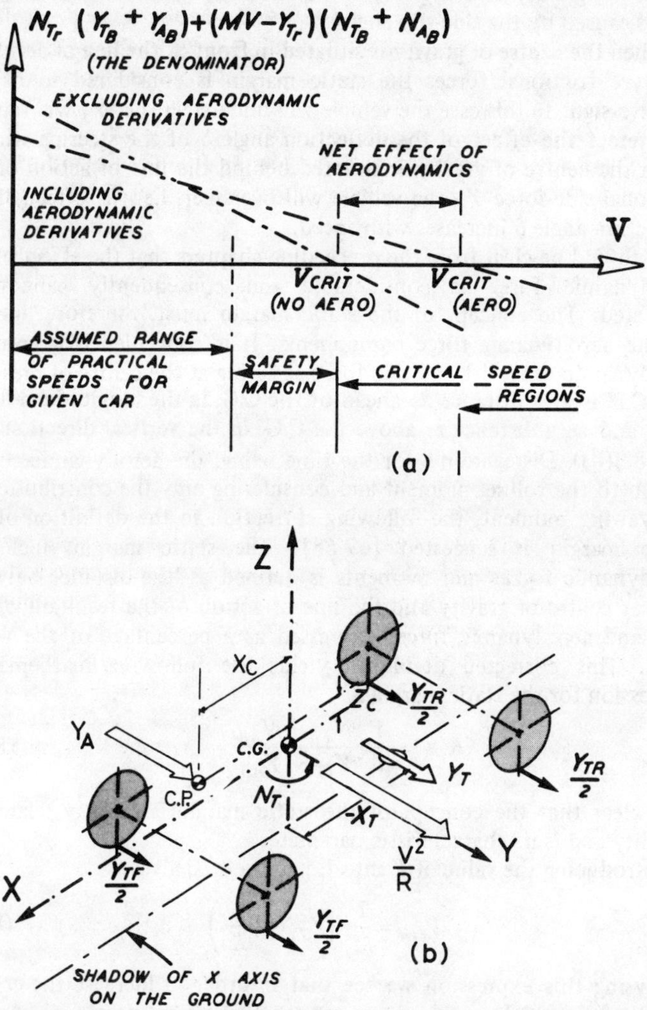

Fig. 8.3. Graphic representation of the critical speed and the safety margin. (b) Forces involved in the definition of static margin

distance was used as a definition of the car yawing moment arm when the problem of stability was treated in the past as an exclusively mechanical problem and the effect of aerodynamic forces was neglected. Thus the concept of the 'static margin' was introduced [62,63,73,75]. The mechanical static margin is defined as 'the distance expressed as a fraction of the wheelbase between the position of the centre of gravity and the line of action of the resultant of the tyre forces caused by the side-slip angle'.

When the centre of gravity is situated in front of the line of action of the tyre frictional forces the static margin is considered to have a positive sign. In this case the vehicle will understeer, i.e. Y_T will tend to counteract the effect of the deflection angle δ of the steering wheels. When the centre of gravity is situated behind the line of action of the frictional side-force Y_T the vehicle will oversteer, i.e. the effect of the deflection angle δ increases with speed.

It should be clear from the preceeding chapters that the effect of the aerodynamic force is considerable and consequently cannot be neglected. The concept of the static margin must, therefore, include all the aerodynamic force components. It is convenient to represent aerodynamic drag, lift and side force as acting at the centre of pressure. The C.P. is at a distance x_c ahead of the C.G. in the longitudinal direction, and at a distance z_c above the C.G. in the vertical direction (see Fig. 8.3(b)). Disregarding, for the time being, the aerodynamic contribution to the rolling moment and considering only the contribution to the yawing moment, the following correction to the definition of the static margin is suggested [67,68]. The static margin, including aerodynamic forces and moments is defined as 'the distance between the car centre of gravity and the line of action of the resultant of the tyre and aerodynamic forces expressed as a percentage of the wheel base'. This corrected definition yields the following mathematical expression for the static margin K:

$$K = -\frac{1}{l}\,\frac{N_{T\beta} + N_{A\beta}}{Y_{T\beta} + Y_{A\beta}} \tag{8.13}$$

It is clear that the concept of the static margin is directly related to stability and is its characteristic parameter.

Introducing the value of K into Equation 8.10 gives

$$V_{crit} = \frac{1}{M}\left(Y_{Tr} + \frac{N_{Tr}}{lK}\right) \tag{8.14}$$

Analysing this expression we see that in order to increase the critical speed of a vehicle we have to reduce the static margin K. As the aerodynamic derivatives are additive to the frictional tyre derivatives it is not directly clear how they affect the value of K. The value of N_A is related to the value of Y_A through the distance x_c. The position of the centre of pressure influences the static margin and as a general rule it is beneficial to reduce, if possible, the distance between the C.P. and the C.G.

8.3 TURNING CHARACTERISTICS

So far we have considered vehicle stability and the road holding behaviour of a car travelling in an almost straight line at different steady speeds, either under the influence of a small sideways wind, or a small front steering wheel deflection. However, driving involves turning when the radius of turning is comparatively small. In discussing the turning motion we have first to simplify the problem by assuming steady turning, i.e. that the speed and the radius are constant. Previous considerations included the radius R and the deflection δ of the front steering wheels and we may conclude that the equations of equilibrium, Equations 8.2 and 8.3, are still basically valid.

Turning involves the generation of a significant centripetal force, and a distinct yaw angle attitude β, both of which effectively introduce a considerable rolling moment Q which, for a small turning radius should not be neglected. Let us now consider the effect of the rolling moment and the rolling deflection angle on the car turning characteristics.

When considering the rolling moment and the rolling angle ϕ we divide the total mass of the car into the 'suspended mass' M_s and the 'unsuspended mass' M_o. Thus the total car mass is

$$M = M_s + M_o.$$

The elastically suspended part is, as we know, free (within limits imposed by the suspension system) to move relative to the rest of the chassis. The rolling moment affecting the sprung mass tilts it so that it reaches the rolling angle ϕ when the opposing moment created by the elastic suspension system is equal to the imposed rolling moment. The reaction moment from the suspension is compensated by the forces between the tyres and the road (Fig. 8.4(a)).

These wheel-road reactions add to or subtract from the natural car weight distribution and thus they vary the individual vertical forces acting between the tyre and the road surface. The rolling, therefore, affects the horizontal wheel-road frictional reactions which are dependent on the vertical loadings. These frictional reactions are important in stability and road holding problems.

Additional vertical wheel-road reactions due to rolling, plus the effect on the relative movement of the car centre of gravity of the

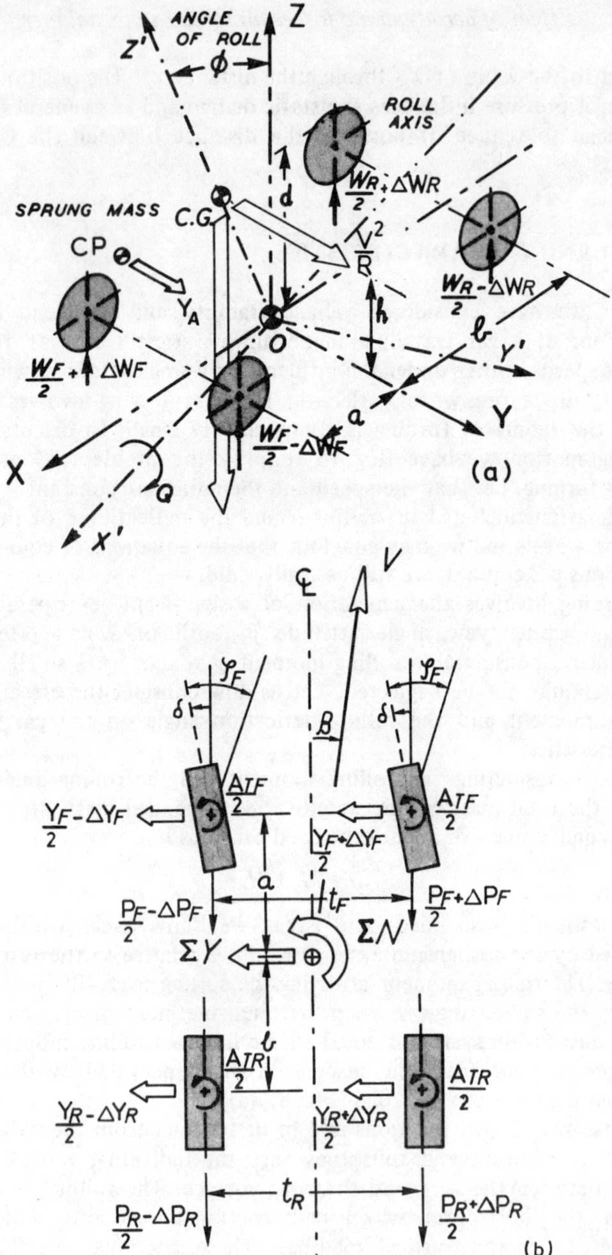

Fig. 8.4. (a) System of rolling axes, sprung mass, and aerodynamic side forces. (b) Tyre rolling drag sideforce and self-aligning torque during turning as a function of car weight

sprung mass due to the rolling deflection angle ϕ, causes the above mentioned differential wheel weight distribution which through the changing conditions of the frictional forces affects the yawing motion of the car.

The total effect of the rolling is therefore expressed in the interference 'coupling' between the rolling and the yawing motion. The effect of rolling on the yawing and side-slip angle is due to:

(1) The wheel load transfer due to the movement of the suspended mass.
(2) The yawing roll steer, due to a change in tyre characteristics.
(3) The change of aerodynamic forces.

The aerodynamic effect is due to the change of aerodynamic characteristics of the vehicle because of its different attitude to the airstream and the direct action of the aerodynamic side force.

When the sprung mass moves sideways during the rolling motion it rotates around the 'rolling axis'. The coupling between the roll and the yaw moments depend on the location of this axis, on the relative position of the centre of gravity of the rolling mass and the position of the aerodynamic centre of pressure [62].

It is assumed here that the reader is familiar with the mechanics of rolling motion and much published literature on the subject exists [74,77,78,79]. Briefly a moving system of axes is considered fixed to the roll axis of the vehicle with its centre at the metacentric intersection (Fig. 8.4(a)). The inertia terms are calculated under a simplifying assumption that the sprung mass has a system of axes which is parallel to the body axes and that the centre of gravity of the non-rolling mass lies on the OX axis.

The purpose of this chapter is to estimate the effect of the superimposed aerodynamic forces on other mechanical effects under steady state conditions. We have therefore to include the rolling effect in the equilibrium condition of forces and moments. Thus we have now an additional equation of equilibrium expressing the fact that for a constant rolling angle ϕ the acting rolling moment (including the aerodynamic force effect) is equal to the deflected suspension reaction torque.

A previous simplifying assumption that the right and left hand wheels behave in the same way is no longer valid and we have to consider the left and right hand side wheels separately. We shall therefore supplement Equations 8.2 and 8.3 with terms containing the rolling effect and evaluate the wheel-load distribution changes due to the tilting of the sprung mass.

The aerodynamic effect of the change of attitude of the sprung part of the car is, in practice, unknown. In further considerations it will be denoted by the derivative

$$\frac{\partial Y_A}{\partial \phi} = Y_{A\phi}$$

and by the yawing moment derivative

$$\frac{\partial N_A}{\partial \phi} = N_{A\phi}.$$

Both these derivatives should be experimentally investigated in the future. Assuming, however, that they have significant values we may express the effect of the aerodynamic side force as $\phi Y_{A\phi}$, and the effect of the yawing moment as $\phi N_{A\phi}$. Since the aerodynamic and mechanical effects are closely connected we have to analyse the effect of the rolling attitude on the roll steer due to the tyres. This is well known and published in the literature relating to the mechanical aspects of car design [60,61,62,65,66,72,77,79]. It is referred to here only because of this close relationship between aerodynamic and mechanical effects.

The variation of the wheel-load causes the variation of the tyre rolling drag *P*,* and the wheel 'self aligning torque' *A*.† Both the self-aligning torque and the rolling drag are a function of the side slip angle. We can write the expressions for the change in rolling drag of the front and rear wheels in the form

$$\Delta P_F = \left(\frac{\partial P}{\partial W}\right)_F \Delta W_F, \qquad \Delta P_R = \left(\frac{\partial P}{\partial W}\right)_R \Delta W_R$$

where $\Delta W_{F,R}$ are the increments in tyre load due to the change of attitude, and the derivatives $(\partial P/\partial W)_{F,R}$ give the slope of the curve of the rolling drag increase with the increase of the load on the tyre. These derivatives have to be found experimentally for a given type of tyre. The tyre side force and the self-aligning torque are also dependent on tyre load in a manner which is to be found in the data supplied by the tyre manufacturers, and from appropriate diagrams we can find the related derivatives. The graphs given in Fig. 8.5(a), (b), (c) and (d) show the side force and self-aligning torque for a typical tyre in terms of the weight acting on the wheel.

* The tyre 'rolling drag' *P* is the resistance of a tyre to rotation caused by the shape of the tyre contact surface on the road. The resultant road reaction passes in front of the wheel axis thus creating a torque resisting the rotation.

† The wheel 'self-aligning torque' *A* is the moment about the vertical axis in the plane of wheel symmetry, created by the pressure and frictional force distribution on the tyre-road contact surface. The tyre yaw angle produces a moment about its vertical axis which tends to reduce this yaw angle.

The load distribution between the wheels on any one axle depends on the height of the centre of gravity of the vehicle and the lateral displacement of the sprung mass due to roll. Neglecting aerodynamic side force at this stage, the balance of moments requires

$$t_F \Delta W_F + t_R \Delta W_R = M_s g \bar{y} - M \frac{\bar{h} V^2}{R}$$

where $t_{F,R}$ are the tracks of the front and rear axles and the increment of load ΔW at one wheel is matched by an equal decrement of load at the other wheel on an axle. To the mechanical effects, we must add the effect on the tyres of the aerodynamic side force Y_A. This force is applied at the centre of pressure, which is usually higher than the centre of gravity. Because, in turning, the side-slip angle β deflects the axis of symmetry of the car into a 'nose-in' attitude on the road, the aerodynamic side force acts in the direction of the centripetal acceleration. This refers to conditions when there is no wind. With an atmospheric wind its net effect could be positive or negative, depending on which side the wind blows at any given part of the turn. The aerodynamic side force creates the aerodynamic rolling moment Q_A which causes a reaction of the suspension and consequently affects the distribution of load on the wheels.

The total vertical load on all the wheels is generally unchanged and so the increase and decrease in load on the wheels on a given axle is largely self-compensating. Thus the differential distribution of wheel loads has self-compensating effects on the rolling drag and on the tyre side forces. The variation of the tyre self-aligning torque with wheel load is a very complex matter generally, as is to be seen from Fig. 8.5(c). However, if we consider the change in this torque from a given value, which arises due to the redistribution of wheel load, we see that the small change associated with an increase of loading on one wheel will be almost exactly compensated by the opposite change due to the decrease in loading on the other wheel at both front and rear axles. However, the effects on yawing moment are additive rather than self-compensating.

The additional yawing moment comes from the moment about the vertical axis of the tyre rolling drag changes due to redistribution of the wheel loading. It consists of 'mechanical' terms due to the attitude of the sprung mass and the turning of the vehicle, and an 'aerodynamic' effect due to the moment of the aerodynamic side force about ground level. The increment in yawing moment is given by

$$\Delta N = P_W \left(M_s g \bar{y} + F_A \bar{d} - \frac{M V^2 \bar{h}}{R} \right) \tag{8.15}$$

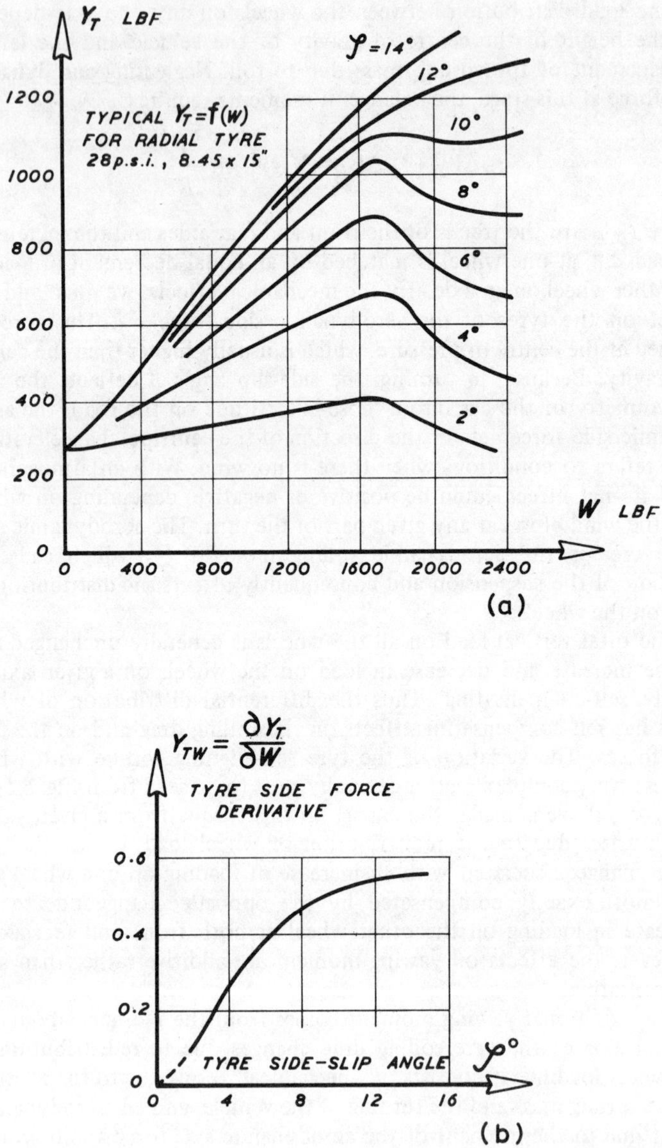

Fig. 8.5. (a) Typical tyre frictional side force at given side-slip angle for different values of the vertical load. (b) Tyre frictional side force derivative with respect to the vertical load as a function of the side-slip angle

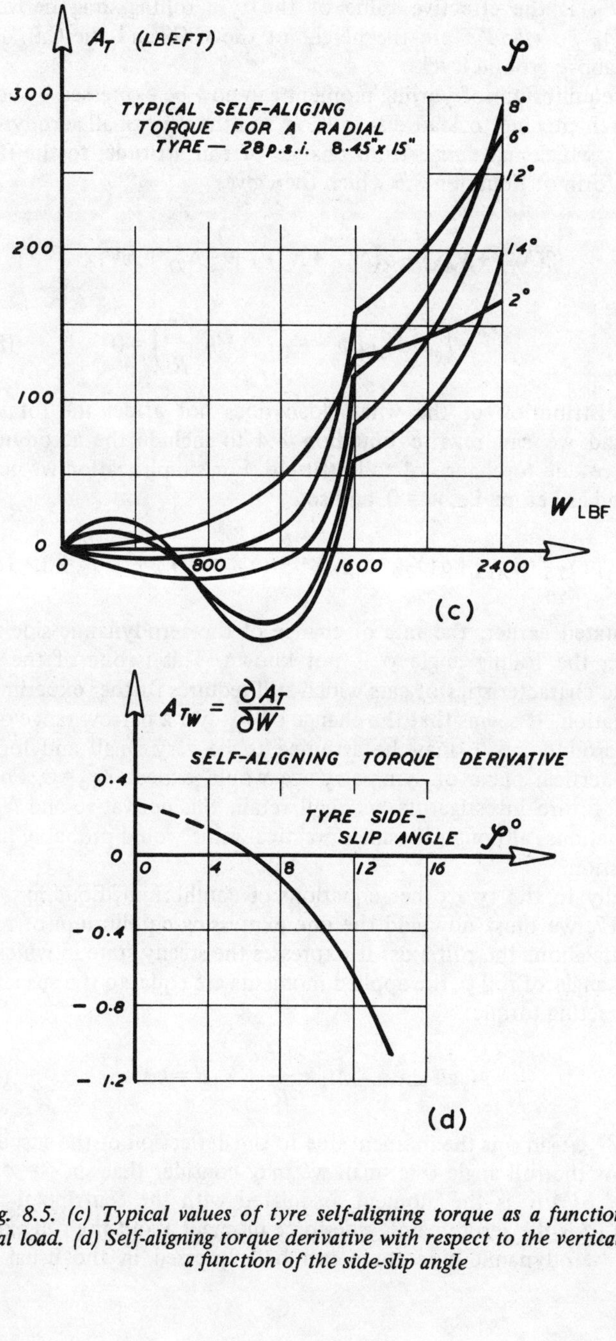

Fig. 8.5. (c) Typical values of tyre self-aligning torque as a function of the vertical load. (d) Self-aligning torque derivative with respect to the vertical load as a function of the side-slip angle

where P_W is the effective value of the tyre rolling drag derivatives $(\partial P/\partial W)_{F,R}$, $\bar{h}$ and $\bar{d}$ are the height of the C.G. and the C.P. of the vehicle above ground level.

The equilibrium of yawing moment can now be expressed by adding the increment due to load distribution and the very small aerodynamic moment which may arise from change of roll attitude, to the fundamental form of Equation 8.5 which then gives

$$\beta(N_{T\beta}+N_{A\beta})+\delta(N_{T\delta}+N_{A\delta})+\frac{V}{R}N_{Tr}+\phi N_{A\phi}$$

$$+P_W\left(M_s gh\phi+F_A\bar{d}-M\bar{h}\frac{V^2}{R}\right)=0 \qquad (8.16)$$

The redistribution of the wheel load does not affect the total side force and we can rewrite Equation 8.4 to include the aerodynamic side force due to change of roll attitude. For simplification we neglect side wind as before, i.e. $w = 0$, and so

$$\beta(Y_{T\beta}+Y_{A\beta})+\delta(Y_{T\delta}+Y_{A\delta})+\frac{V}{R}(Y_{Tr}-MV)+\phi Y_{A\phi}=0. \qquad (8.17)$$

As stated earlier, the rate of change of the aerodynamic side force Y_A with the rolling angle ϕ is not known. This is one of the aerodynamic characteristics of cars which still requires further experimental investigation. It seems that the change of Y_A over a narrow range of roll angle (around $\pm 5°$), may be assumed to be very small and for cars with a vertical plane of symmetry we would expect $Y_{A\phi}=0$. For the sake of future investigators we shall retain this derivative and $N_{A\phi}$ in our equations, although its small practical value would probably justify its omission.

Finally to the two other equations of equilibrium, Equations 8.16 and 8.17, we must now add the one expressing equilibrium of rolling moments about the roll axis. It expresses the steady state in which, for a given angle of roll ϕ, the applied moments are equal to the suspension counteracting torque:

$$M_s gh\sin\phi-M_s h\frac{V^2}{R}+Y_A d=k\phi \qquad (8.18)$$

where $M_s gh \sin\phi$ is the moment due to the deflection of the suspended mass. As the roll angle ϕ is small we may consider that $\sin\phi \simeq \phi$. The term $M_s hV^2/R$ is the moment associated with the centripetal force, while $Y_A d$ is the aerodynamic side-force moment about the roll axis.

The aerodynamic side force can be expressed in the usual aerodynamic form and as it is present here due to the side-slip angle β we may express it in the form

$$Y_A = Y_{A\beta}\beta$$

Thus we can write Equation 8.18 expressing the equilibrium of the moments about the roll axis as follows:

$$\beta Y_{A\beta} d - \frac{V}{R} M_s V h + \phi(M_s g h - k) = 0. \qquad (8.19)$$

In steady state motion the action of the shock absorbers is negligible and only the torsional stiffness of the elastic suspension k is considered.

The three equations, 8.16, 8.17 and 8.19 allow us to calculate the yaw angle β and the car turning radius R in terms of the front wheel steering deflection angle δ. Calculating the roll angle from Equation 8.19 and substituting in Equation 8.17 we can calculate the side-slip angle β in terms of the angle δ and radius R

$$\beta = -\frac{\delta(Y_{T\delta} + Y_{A\delta}) + \frac{V}{R}(Y_{Tr} - MV - Y_{A\phi}B)}{Y_{T\beta} + Y_{A\beta} + Y_{A\phi}A}$$

where

$$A = \frac{Y_{A\beta} d}{k - M_s g h} \quad \text{and} \quad B = \frac{M_s V h}{k - M_s g h}$$

in these equations. Since we can write the roll angle as

$$\phi = A\beta - B\frac{V}{R}$$

and if we assume that the centre of gravity of the unsprung mass lies close to the roll axis, near axle height, we can write the increment in yawing moment given by Equation 8.15 as

$$\Delta N = P_W\left\{(Y_{A\beta}\bar{d} + AM_s g h)\,\beta - (M\bar{h}V + BM_s g h)\frac{V}{R}\right\} = P_W\left(G\beta - H\frac{V}{R}\right)$$

where G and H depend on the mass distribution, geometry, aerodynamic and suspension characteristics of the vehicle.

Substituting this expression into Equation 8.16 and then eliminating V/R and β in turn from Equations 8.16 and 8.17 gives the side-slip angle β in terms of the steering angle δ as

$$\beta = -\delta \frac{(Y_{T\delta} + Y_{A\delta})(N_{Tr} - HP_W - BN_{A\phi}) - (N_{T\delta} + N_{A\delta})(Y_{Tr} - MV - Y_{A\phi}B)}{(N_{Tr} - HP_W - BN_{A\phi})(Y_{T\beta} + Y_{A\beta} + Y_{A\phi}A) - (Y_{Tr} - MV - Y_{A\phi}B)(N_{T\beta} + N_{A\beta} + GP_W + AN_{A\phi})} \tag{8.20}$$

and the turning radius R in terms of the steering angle δ as

$$\frac{V}{R} = -\delta \frac{(N_{T\delta} + N_{A\delta})(Y_{T\beta} + Y_{A\beta} + AY_{A\phi}) - (Y_{T\delta} + Y_{A\delta})(N_{T\beta} + N_{A\beta} + GP_W + AN_{A\phi})}{(N_{Tr} - HP_W - BN_{A\phi})(Y_{T\beta} + Y_{A\beta} + AY_{A\phi}) - (Y_{Tr} - MV - BY_{A\phi})(N_{T\beta} + N_{A\beta} + GP_W + AN_{A\phi})} \tag{8.21}$$

On setting the terms A, B, G and H to zero, Equation 8.20 simplifies to Equation 8.11 and Equation 8.21 simplifies to Equation 8.12 in the primary analysis of the steady state without load redistribution and rolling effects. The forms resulting when $Y_{A\phi} = N_{A\phi} = 0$ are easily determined. Some of the derivatives appearing in Equations 8.20 and 8.21 require practical investigation. Clearly the change of the yaw moment due to the redistribution of the load on the tyres due to turning and change of roll attitude is quite considerable, is generally adverse and is consequently of importance.

The present use of modern cars on fast roads and in windswept regions, driving in all weather conditions, necessitates a further extension of the investigations into the aerodynamic effects on car stability, exploring even the exceptional circumstances in which car instability can occur. Tyre derivatives are usually quoted for dry road surface conditions but on wet roads, or in icy weather, the value of these derivatives can change drastically and then the aerodynamic effects could very easily predominate.

When we consider the appropriate values of car derivatives for the characteristic dimensions of an average saloon we may assume that the values of A and B above are of the order of 0.5 and 3.0 respectively and that these values are slightly dependent on the vehicle speed. Assuming a linear variation of A and B with car speed, we can estimate their value at any other speed by interpolation between two known values. Thus an estimation of the rolling effect on the critical velocity of a car can be made. This critical velocity, at which the vehicle reaches a state of uncontrollable spin is obtained, as before, by equating the denominator of Equation 8.21 to zero and calculating the velocity

$$V_{\text{crit}} = \frac{Y_{Tr} - BY_{A\phi}}{M} - \frac{(N_{Tr} - HP_W - BN_{A\phi})(Y_{T\beta} + Y_{A\beta} + AY_{A\phi})}{M(N_{T\beta} + N_{A\beta} + GP_W + AN_{A\phi})} \tag{8.22}$$

If we set the values of A, B, G and H to zero, this equation reduces to Equation 8.10 which gives the critical speed while neglecting rolling effects.

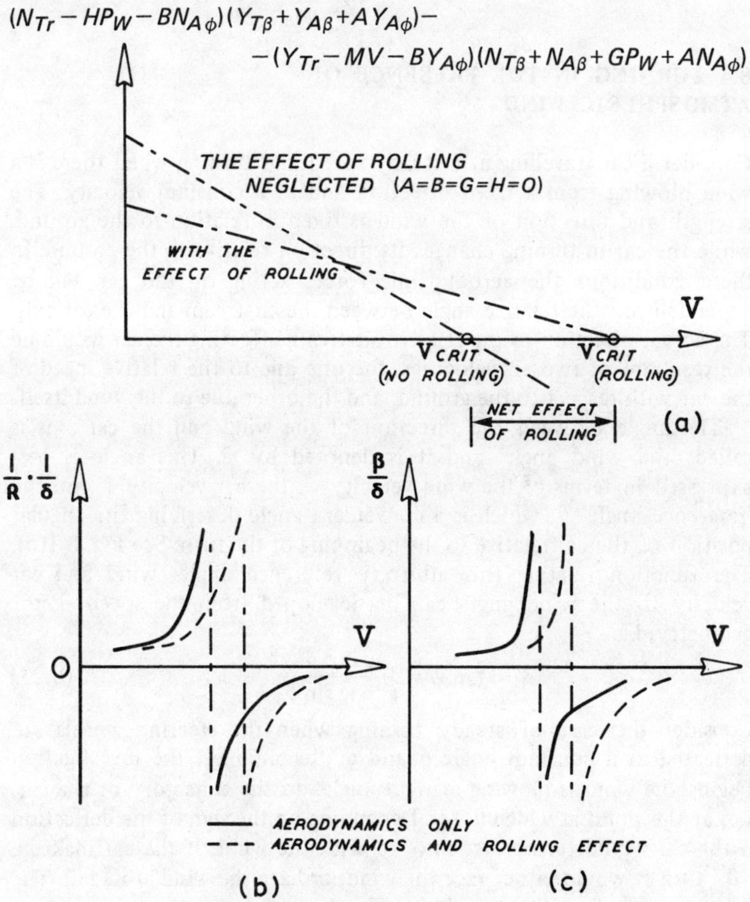

Fig. 8.6. (a) The effect of coupling between yawing and rolling moments on critical speed. (b) Effect of rolling attitude on the ratio of the steering wheels deflection angle to car turning radius as a function of speed and (c) ratio of the yaw to steering angle as a function of speed

An analysis of Equation 8.22 shows that the effect of the rolling attitude tends to increase the critical speed, which means that the exclusion of the rolling moments, and consequently of the 'coupling effect' between the rolling and yawing motions, leads to a pessimistic estimation of car stability. The effect of a rolling-yawing coupling is shown in Fig. 8.6(a). This also applies to Equations 8.20 and 8.21. The effect of coupling between rolling and yawing increases the speed at which either δ/R or β/δ tends to infinity, i.e. the car becomes uncontrollable. This is shown in Fig. 8.6(b) and (c).

8.4 TURNING IN THE PRESENCE OF ATMOSPHERIC WIND

Consider a car travelling at a steady speed and turning when there is a wind blowing from a fixed direction and at a constant velocity. The strength and direction of the wind is fixed in relation to the ground, while the car in turning changes its direction relative to the ground. In these conditions the aerodynamic forces acting on the car will be dependent on the relative angle between the airstream and the car axis. The actual direction of the relative airstream affecting the car would be the resultant of two air velocities, the one due to the relative speed of the car with respect to the ground, and the other due to the wind itself.

The angle between the direction of the wind and the car axis is called 'the wind angle' and it is denoted by ψ. This angle is best expressed in terms of the wind velocity w, the car velocity V and the 'reference angle' θ, which is a convenient angle describing the angular position of the car relative to the beginning of the turn. See Fig. 8.1(c). The function relating this arbitrary reference angle, wind and car velocity to the wind angle can be developed from the speed vector geometry, i.e.

$$\tan \psi = \frac{w \cos \theta}{V \pm w \sin \theta} \tag{8.23}$$

Consider the case of steady turning when the steering wheels are deflected at a constant angle δ and at the moment the investigation begins the wind is blowing at right angles to the trajectory of the car, i.e. at the point at which $\theta = 0$. Depending on the sign of the deflection δ the car would turn either into, or with the wind. If the car makes a 90° turn it would either face the wind, or have the wind at its tail. The subsequent trajectories would be different in each case.

The relative direction of the side wind changes during turning. This change is of a cyclic character, i.e. from $\tan \psi = w/V$ to $\tan \psi = 0$. Each position of the car during the process of turning is considered as 'steady', i.e. we disregard any acceleration effects. As the relative angle between the resultant airstream is now $(\beta \pm \psi)$ we have to rewrite Equations 8.16 and 8.17, together with Equation 8.19, to include this correction, i.e. the aerodynamic derivatives $Y_{A\beta}$ and $N_{A\beta}$ are multiplied by the sum of the angles $(\beta \pm \psi)$. Eliminating the rolling angle ϕ and the yaw angle β and calculating the turning radius R in terms of the angle of deflection δ and the wind angle ψ we obtain

$$\frac{1}{R} = \left(\frac{1}{R}\right)_{w=0} + \frac{\psi}{V} \cdot \frac{E}{n\Delta} \tag{8.24}$$

In this expression the value of $(1/R)_{W=0}$ is in fact the whole of Equation 8.21, i.e. it is the value of the turning radius in the 'no wind' case, where E denotes the following function

$$E = (Y_{A\beta} + AY_{A\phi})(N_{T\beta} + N_{A\beta} + GP_W + AN_{A\phi}) - (Y_{T\beta} + Y_{A\beta} + AY_{A\phi}) \times (N_{A\beta} + GP_W + AN_{A\phi}) \tag{8.25}$$

The symbol $n\Delta$ denotes the denominator of Equation 8.21. These formulae, and especially Equation 8.24, show that the turning radius changes during the turn due to the wind effect. These changes are different for turning into the wind and for turning with the wind. The variability of the turning trajectory for these two cases is shown in Fig. 8.7(a). The diagram gives values of the ratio of comparative radii $R/R_{W=0}$, with and without the wind effect for a fixed steering wheel deflection δ at a varying reference angle θ during a 90° turn. This diagram was based on numerical data available for two typical racing cars (a classic Formula I with large exposed wheels, thus including the derivatives $Y_{A\delta}$ and $N_{A\delta}$, and a fully streamlined G.T. type).

It is of interest to note that the aerodynamically better styled car (i.e. fully streamlined) is, in effect, much more sensitive to the wind in turning than the less aerodynamically styled car. The wind effect when turning into the wind is much smaller, resulting in a marked increase of the turning radius for at least two thirds of the turn, while in the last stages of the turn the turning radius increase is gradually reduced to zero. Turning with the tail wind produces a different effect. The turning radius is decreased during the first stage of the turning. This decrease is comparatively greater than the increase of R when turning into the wind. The lengthening is more gradual in the final stages of the turn. The dotted line curve in Fig. 8.7(a) shows the estimated effect of the coupling between the rolling and the yawing moments. The effect of the coupling improves the ratio $R/R_{W=0}$ by approximately 2% when turning into the wind and by about 5% when turning with the wind. When turning into the wind the turning radius tends to decrease, while in turning with the wind the radius increases. The effect of the actual

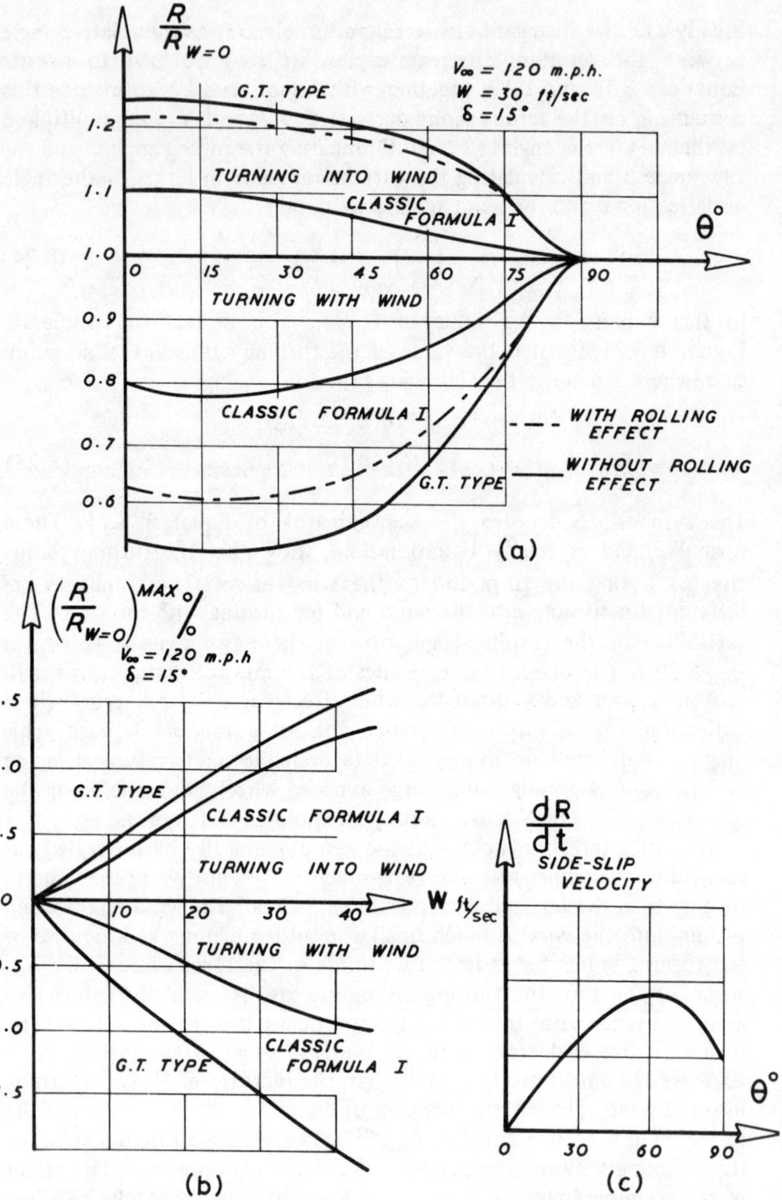

Fig. 8.7. (a) The side wind effect on turning radius changing with the direction of turning–with or into the wind. (b) Maximum change in turning radius as a function of the side wind speed. (c) Side-slip (transverse) velocity during a 90° turn

speed of the side wind on the turning radius is shown in Fig. 8.7(b).

We have to remember that these considerations were made assuming zero deflection of the steering wheels. In practice the necessary correction to the turning radius is carried out instinctively by the driver who by viewing the road in front steers along a pre-conceived line.

Figure 8.7(a) shows distinctly that the turning radius varies with the progress of the turn. The rate of radius change, i.e. dR/dt is nothing else but the speed of the 'side-slip' of the car in a radial direction. Denoting this side-slip velocity by U, we can calculate it by differentiating Equation 8.24 with respect to time. This operation is complex and the resultant equation rather complicated. Introducing tentative values for a typical car we can estimate the side-slip velocity U, during the turn, as a function of the reference angle θ. Results of such an estimation are shown in Fig. 8.7(c).

It is interesting to note that the side-slip velocity increases rapidly at the beginning of a turn into the wind and reaches its maximum at approximately 2/3 of the 90° turn, i.e. at $\theta = 60°$. It is, as many drivers have experienced, the most dangerous moment in turning when the side-slip may become uncontrollable on a patch of oil or on a wet road. The apparent anomaly of zero side-slip velocity at the beginning of the turn may be explained by the original assumption that the turn is instantaneously begun when $\theta = 0$. This means that the steering front wheels are hypothetically deflected from zero to full deflection in zero time at the beginning of the turn and also we assumed that the wind was applied suddenly at this moment. This last assumption could be realised when the vehicle runs into a side wind from a still atmosphere. In any case these simplifying assumptions are only hypothetical and are simply a means of providing an approximate analysis. In practice any sudden change of the conditions of motion would be followed by a short or in some cases even a prolonged period of transient motion in which most parameters of motion would be 'time dependent'. This involves mass inertia effects and accelerations. This kind of motion is then of a primarily dynamic character.

Chapter 9

DYNAMIC RESPONSE TO TRANSIENT AERODYNAMIC FORCES

9.1 INTRODUCTION

We discussed in the preceeding chapters steady state motion. By assuming this we simplify the problem, reducing it to an analysis of the equilibrium of constant value forces and moments affecting a car. These simplifying assumptions were then complicated by the gradual addition of the coupling effect between the rolling and the yawing moments, and finally by the addition of the wind effect in turning.

Useful as this simplification is in the study of the fundamental principles of steady state car stability and the analysis of the aerodynamic effect on it, it does not effectively represent actual road conditions. In practice driving on an ordinary road rarely corresponds to steady state motion. This idealised situation in which all the forces and moments are in equilibrium has to be upset now and again, either by driver action, necessitated by road conditions, or by a sudden change of the external airstream due to a wind gust or the turbulent wake caused by the wind and ground configuration, or by other traffic on the road.

Every experienced driver knows the effect a strong cross-wind has when overtaking say a large lorry—the car seems to be 'sucked in' towards the lorry and often a quick steering correction is necessary. Racing drivers well know the technique of overtaking at speed when they drive within the wake-region of the car in front. This helps to gain additional momentum which is needed in the manoeuvre of overtaking. Nation-wide radio warnings to drivers of wind gusts affecting certain sections of motorways are commonplace today.

All these phenomena are basically of an aerodynamic character. More powerful and faster cars, a modern road system, and denser traffic, all these factors make the problems of car safety more pressing and important. The correct understanding of aerodynamic phenomena is now essential not only to car stylists and designers, but maybe even more importantly, to car users. How important these aerodynamic problems are can be estimated from the following example which occurred some years ago when the racing world was shaken by a fatal accident in which a famous driver lost his life. There was no official explanation of the accident apart from a short statement that the 'design and mechanical working of the car was not to blame'. However it is interesting to speculate about the possible reason for the accident based on details of the incident reconstructed from conversations with witnesses. It happened on a famous racing track. The car was travelling at about 180 m.p.h. prior to entering a slightly bending right hand turn. The road ran across open ground and the car was subjected to cross-wind conditions. Down the bend the road ran into a thick fir forest which acted as a wind shield. On the open, unsheltered road the driver maintained a 'steady-state' condition with the forces and moments, including the effect of a constant cross-wind, in equilibrium. Racing technique induced the driver to run very near the inner edge of the road as it was turning right. The deflection of the steering wheels was such that it coincided with the equilibrium conditions. When the car suddenly passed from the wind swept road into the sheltered region behind the fir forest the external aerodynamic forces underwent a rapid change due to the disappearance of the wind component. The dynamic forces acting on the car were unbalanced and the steering wheels deflection angle was not correct. Before the driver could react correctly at this speed his right hand side wheels had run into the soft shoulder of the road. Regaining control over the car in the circumstances was impossible. The deflection of the wheels in this case was quite small, as the skid marks indicated.

Modern motorways, with dense and comparatively fast traffic, in conditions of atmospheric wind often create conditions in which such transient motion may lead to a change in trajectory which, however small, in some cases could cause an accident. It is interesting to conjecture, in the absence of reliable statistics on this matter, how many so-called 'unexplained' accidents are in fact caused by such a sequence of events. Consider for example the comparatively frequent type of motorway accident, i.e. the unexpected cross-over of the central reservation and a collision with an oncoming car. The danger of this type of accident has been reduced, at great cost, by the installation of central crash barriers. Mechanically minded statisticians clearly suspect the tyres which when suddenly deflated cause a rapid loss of control. It is nevertheless rather curious that the front wheel tyres burst more frequently on the right hand rather than on the left hand side. It may be suspected by aerodynamicists that the external aerodynamic forces could, at least in a number of cases, have caused the right front wheel to 'climb-up' unexpectedly on to the central curb, leading to a loss of control, or tyre burst. A small change in the car trajectory (sometimes quite unnoticed by the driver) could, when the car is driven too near the curb, cause such behaviour.

There are many reasons for an impulsive change in the external aerodynamic forces and its effect leads to an oscillatory motion of the car which often requires steering correction. It could be a gust of cross-wind, a sheltered region behind a building, or the wake behind other vehicles. It would be very instructive to study accident statistics which include wind conditions or other relevant aerodynamic phenomena, which leads us to the problems of transient car motion.

Aerodynamic phenomena should be looked upon from two different points of view. We can consider the problem of the atmospheric wind and the kind of ground configurations to be avoided to protect traffic from wind gusts or on the other hand we can consider the problem of the variable aerodynamic force which can abruptly change driving conditions and requires quick and correct steering responses from the driver.

Both these approaches should be made and are essential to improve car safety. It is then the purpose of this chapter to present these problems in a form which is readily comprehensible but which also outlines the theory involved. This approach may be useful in indicating a prospective field of study and the extent of the problems still to be solved. It does not offer any ready solutions.

As this book deals primarily with car aerodynamics the problem of the wind and its effect on the design of road systems is omitted and we concentrate on car dynamics affected by transient aerodynamic forces.

We shall assume that the external aerodynamic force is not constant but varies in time and can vary both in magnitude and direction. This variability of the aerodynamic force introduces mass inertia forces due to acceleration. Thus the phenomena described are of a dynamic nature and consequently the car motion becomes time dependent, i.e. transient. The mathematical description of such transient motion is in the form of equations of motion in which all forces and moments include the effects of acceleration and inertia.

The sudden change of the aerodynamic force, acting externally, can cause a vehicle to accelerate or decelerate and to change its attitude on the road. This state of motion is called 'transient' and can lead to variations of the yawing or rolling attitudes, connected with variations in the turning radius, which have to be corrected by driver action. Thus the transient state requires a driver response in changing the deflection angle of the front steering wheels. In practice, as we previously discussed, cars have no automatic way of correcting changing attitudes, and therefore, driver response is necessary. This is not completely true as, in practice, some vehicles have a small degree of automatic stability, requiring only a degree of steering response from the driver. This response is required only when the effect of the aerodynamic force exceeds the limits with which the automatic stability operates.

The average person has a response time of approximately 1/16 seconds. This is the time which elapses before the senses alert and the muscles begin to operate. Therefore, when the initial impulse of the external aerodynamic force begins to change the car attitude the process is well advanced before the driver can possibly begin to react. There is a 'time-lag' between the beginning of the initial impulse and the reaction. The resultant car motion is therefore basically of an oscillatory character. The characteristic of this oscillatory motion is called the 'dynamic stability' of the car.

If a car has good damping characteristics and the oscillations quickly cease we say that the car has 'good dynamic stability'. The changes in trajectory and yaw attitude due to the action of transient aerodynamic forces can be corrected by the driver through the steering mechanism. However the system has a 'built-in' delay due to the reaction time of the driver and the response time of the steering mechanism. This superposition of the vehicle response (due to the aerodynamic force) and driver response (due to driver correction) results in a motion which is oscillatory in character. When this motion develops into large and sometimes increasingly oscillatory deviations we say that the car is 'dynamically unstable'.

Those who are familiar with the aeronautical term 'dynamic stability' will see that the meaning of this term is comparable in both aircraft and road vehicle technology. In aeronautics the same term refers only to the damping oscillatory motion of the aircraft as its characteristics exclude the pilot's response, i.e. only the behaviour of the aircraft with its controls frozen is considered, which may or may not dampen the oscillations.

In car dynamic stability we cannot avoid driver response as it plays a crucial part in the whole process of damping the oscillations. We are not enlarging here on the dynamics of steering, which is a problem in itself, but we discuss it only insofar as it is involved with the aerodynamic effect on the dynamic stability. Figure 9.1(a) shows changes with respect to time of the side-slip angle due to, first the sudden increase of the aerodynamic side force ΔY_A and then, independently, to the steering impulse $\Delta\delta$. This diagram illustrates two independent effects which in a real case would be superimposed within the time-lag explained before. The variations of the yaw angle β caused by the transient aerodynamic force are actually slower than those due to the steering corrections.

A school of thought exists which advocates that the car designer should try to make the vehicle 100% stable, i.e. able to correct automatically (with the steering wheels frozen) for any external impulsive force. This demand has proved to be impractical and at present prevailing opinion is that a certain degree of automatic

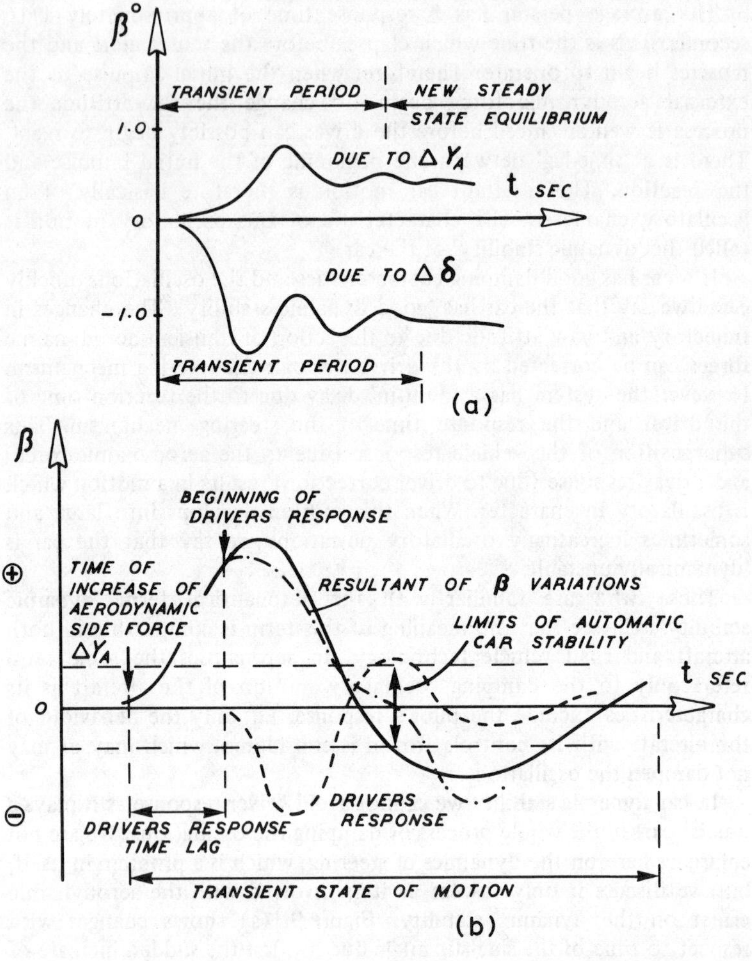

Fig. 9.1. (a) Periodic variations of yaw angle caused by transient aerodynamic forces or by impulses due to steering wheels. (b) Periodic pattern of yaw angle variation after a single aerodynamic force impulse followed by correcting response of driver

corrective response is necessary, but the main correction for large deviations should be made by driver action (Fig. 9.1(b)).

In such a system drivers use the front wheel angle setting δ for general 'long term' steering and 'short term' variations of the steering wheel deflection $\Delta\delta$ for the dynamic stabilisation of the car. This is therefore, in principle, a logical sequence of responses which is triggered by the sudden increase (or decrease) of the external aerodynamic force.

Investigation of transient state car motion is now in progress in research centres throughout the world [63,73,74,76]. This research is by no means conclusive. In view of the present state of knowledge of this subject this chapter can only be of an introductory nature.

9.2 HISTORICAL DEVELOPMENT OF CAR DYNAMIC STABILITY

The development of automobile technology shows that the problems of vehicle handling characteristics and responses to steering were first studied in 1935 (Evans [77]) and later in 1946 (Olley [78]) when a fundamental theory of vehicle handling was elaborated. These studies were primarily concerned with the mechanics of steering and with the behaviour of a car in dynamic oscillation. A serious study of the theory and analysis of response to external impulses began later, about 1950, so that this subject is comparatively recent.

After the Second World War the investigation of dynamic stability was concentrated in creating a mathematical analogue system to represent the dynamics of vehicle response to steering and to external aerodynamic forces. This coincided with the general use of computers. A mathematical analogue system was developed during 1954 to 1956 [62,63,79], and followed later by other workers [52,80]. Preparing computer programmes for the investigation of dynamic stability was completed in 1961 [81], which work was further improved in the U.K. in 1963 [66].

These papers and this research work were mainly concerned with the steering response. These studies, both theoretical and experimental, considered exclusively the steering mechanism and inertia force. The high state of perfection achieved has thus prepared the ground for an analysis of the addition of aerodynamic forces on vehicle stability characteristics.

9.3 TRANSIENT AERODYNAMIC FORCES

Previously we have been concerned with a steady relative airstream velocity and a steady state aerodynamic force acting at the centre of pressure of a car. Now we consider the airstream relative velocity varying with time. The changes of relative air velocity can be either positive or negative, i.e. the velocity can either increase or decrease over a short period of time. It is important to realise that this airstream velocity change is brought about by external causes, i.e. it does not originate from the vehicle which has a constant velocity with respect to the ground. Thus the air velocity change is due to the atmospheric wind, or some other external cause, such as a turbulent wake, which a car might encounter.

It is not the purpose of this book to discuss the atmospheric wind in detail and the interested reader will find much published literature on this subject, for example [18,72,83]. The atmospheric wind is especially turbulent near the ground where it tends to form eddies of different sizes. On windy days large structures, or buildings near the road, and the natural ground configuration can create a turbulent wake downwind, often extending across the road. Overtaking stationary or slower vehicles also involves running into turbulent wakes.

Some motorways are elevated above the surrounding countryside and the ground configuration helps to amplify the local wind velocity. For example, mountain passes or the sides of valleys are often places where the local wind velocity can reach up to ten times the normal atmospheric wind velocity. Radio warnings to motorists about such local conditions are now commonplace. It is therefore of considerable importance to be able to estimate the strength and duration of a wind gust and to investigate its effect on the response of a car.

Wind gusts

The wind gust is a natural phenomenon and occurs in a random manner. The recording of random natural gusts are made regularly by meteorologists. Such a recording is shown schematically in Fig. 9.2. The wind velocity is measured by sensitive anemometers and the inherent turbulence in the airstream is denoted by the random oscillations of the instruments. The main characteristic of such a 'wind gust' is a sudden increase in wind speed which takes place in a finite time. An infinitely sharp gust is physically impossible and we have a time-distance over which the speed builds-up either from zero, or from a small initial velocity. Investigation of the natural wind gust seems to indicate that the velocity change has a rather sinusoidal character but often, especially in the case of large and sharp gusts of a long duration, the velocity change is linear and can be described as a flat step with rounded eddies (Fig. 9.2).

It is important to estimate the duration of a gust [82,83] and its size relative to the car. The ratio of the gust duration to the size of a vehicle is referred to by different authors as the 'gust volume'. This ratio gives us a general idea of how the gust can influence the motion of a vehicle. When the gust duration and its dimensions are much smaller than the size of the vehicle it will run through without reacting to the momentarily changed conditions. On the other hand the vehicle, when the gust duration is sufficiently long and its size many times greater

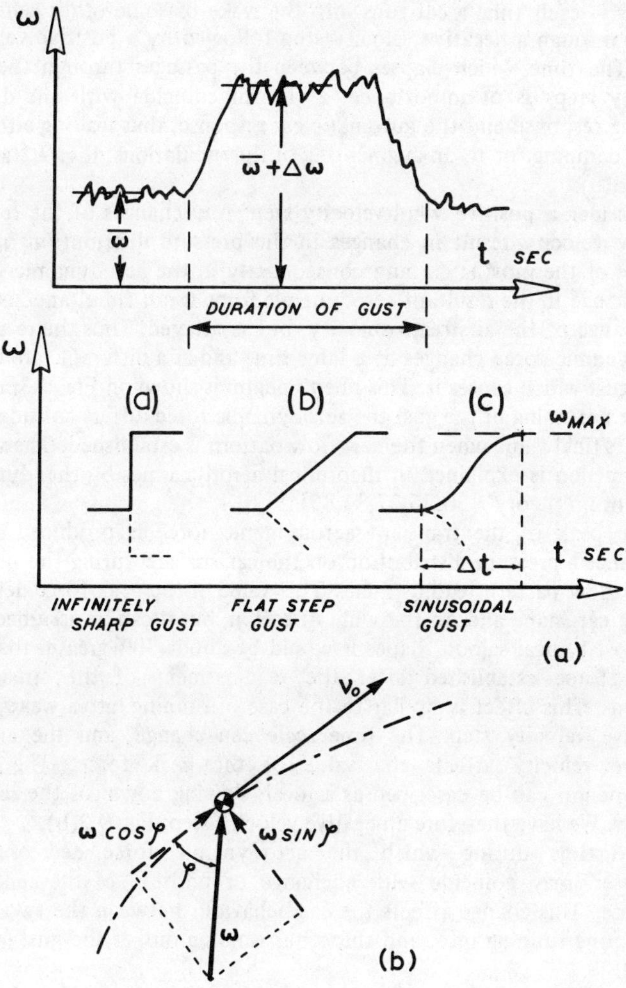

Fig. 9.2 (a) Typical measurement of wind velocity and a simplified type of wind velocity step during a wind gust. (b) The wind angle and wind components relative to direction of motion of car

than the size of the vehicle, will be subjected to the changed conditions during a sufficiently long period of time to be dynamically affected.

The analytical study of the transient aerodynamic force effect has to concentrate then on the study of how an external 'velocity step' affects the car motion. Each gust of wind is composed of a positive velocity step followed by the gust duration and then by a negative velocity step.

Similarly, each time a car runs into the wake of some other vehicle it passes through a negative velocity step followed by a positive velocity step. The time which elapses between the passages through the two velocity steps is of importance, as it can coincide with the drivers steering response and the automatic car response, thus leading either to better damping, or to an augmenting of the oscillations, Fig. 9.2(a) and Fig. 9.3(b).

Consider a positive wind velocity step. Any changes of the relative airflow velocity result in changes in the pressure distribution on the surface of the moving car and consequently in the aerodynamic force. The change in the resultant aerodynamic force is not simultaneous with the change of the airstream velocity, but is delayed. Thus the resultant aerodynamic force changes at a later time and at a different rate to the wind gust which causes it. This phenomenon is shown in Fig. 9.3(a). At the beginning of the gust the aerodynamic force differs considerably from its final value when the new flow pattern is established. The whole phenomenon is explained in theoretical aerodynamics by the 'dynamic momentum theory' [20,26,27,84,85].

The peak of the transient aerodynamic force is produced by an unbalanced pressure distribution on the car surface during the process of the flow pattern readjustment. This value of the peak force depends on the car shape and on the wind direction, but it can be assumed that for most typical saloon shapes it would be about 30% greater than the actual force established after the readjustment of the streamline patterns. This effect is similar to the case of running into a wake, i.e. a negative velocity step. The time scale can change, and the air–car relative velocity affects the value of the peak force. The wake phenonenon can be described as a local 'slowing down' of the relative airflow. We have therefore a negative velocity step, Fig. 9.3(b).

The time during which the aerodynamic force decreases, or increases, may coincide with a change of position of the centre of pressure. This change affects the car behaviour between the two force peaks, one running into, and the other running out of the gust (or the wake).

Consider a car moving relative to the ground at a speed V_o, when the wind velocity also relative to the ground is directed at an angle ψ to the line tangential to the car trajectory (Fig. 9.2(b)). The wind velocity vector can be resolved into the normal and parallel component vectors, relative to the trajectory. The step in the intensity of the wind can be considered either as sinusoidal or as a flat step (Fig. 9.2(a)). Thus the variable wind velocity can be expressed either by

$$w = w_o \left\{ 1 + \frac{w_{max} - w_o}{w_o} \sin \left(\frac{\pi t}{2\Delta t} \right) \right\} \qquad (9.1(a))$$

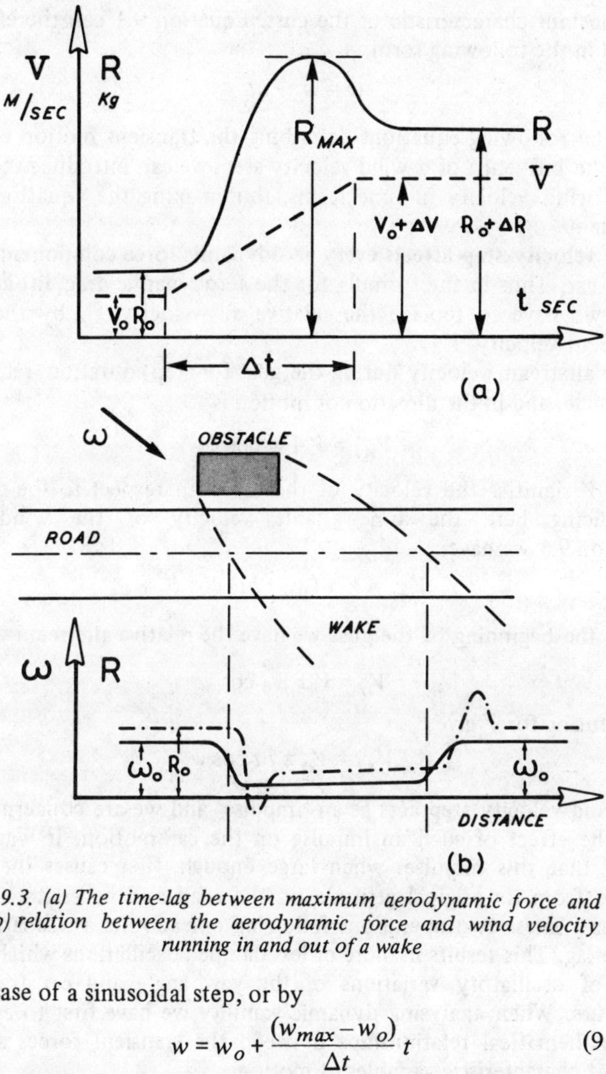

Fig. 9.3. (a) The time-lag between maximum aerodynamic force and a wind gust; (b) relation between the aerodynamic force and wind velocity during running in and out of a wake

in the case of a sinusoidal step, or by

$$w = w_o + \frac{(w_{max} - w_o)}{\Delta t}\, t \qquad (9.1(b))$$

in the case of a linear step. The symbol w_o is the initial wind velocity, Δt is the duration of the wind gust, $t/2\Delta t$ is the time coefficient and w_{max} is the maximum wind velocity.

For convenience and simplicity we may assume a linear characteristic and consider the value of the ratio

$$\frac{w_{max} - w_o}{\Delta t} = b$$

as a constant characteristic of the gust. Equation 9.1 can therefore be written in the following form

$$w = w_o + bt.$$

In the following equations describing the transient motion of a car under the influence of a wind velocity step, we can introduce the value of the wind velocity in time terms, thus making the equation 'time dependent'.

The velocity step affects every aerodynamic force component acting on the car. Thus in the formula for the aerodynamic drag, lift and side force, we have to replace the relative air velocity V_A by the 'time dependent velocity' V_{At}.

The airstream velocity during the gust (or step) duration, relative to the vehicle, and in the direction of motion is

$$V_{At} = V \mp w \cos \psi$$

where V signifies the velocity of the car with respect to the ground. Introducing here the time-variable velocity of the wind from Equation 9.1 we have

$$V_{At} = V \pm (w_o + bt) \cos \psi$$

and at the beginning of the gust we have the relative airstream velocity

$$V_o = V \pm w_o \cos \psi$$

Substituting for V gives

$$V_{At} = V_o \pm bt \cos \psi \tag{9.2}$$

The wind velocity step acts as an 'impulse' and we are concerned here with the effect of such an impulse on the car motion. It was stated earlier that this impulse, when large enough, first causes the car to deviate from its original attitude on the road and necessitates driver response. These two operations following one another are separated by a time lag. This results in more or less damped oscillations which take a form of oscillatory variations of the yaw angle and car trajectory curvature. When analysing dynamic stability we have first to establish the mathematical relationships between the transient forces and the relevant characteristic variables of motion.

9.4 EQUATIONS OF TRANSIENT STATE MOTION

The description and consequent analysis of transient car motion can be carried out only when we develop mathematical formulae which relate all the relevant variables and car characteristic values. There are, as in all these problems of car motion, two possible approaches to the problem. Some authors suggest (for example Hales [65]), that first all six 'degrees of freedom' of motion should be expressed mathematically* and the least important terms in these equations should be eliminated to simplify them as much as possible. This method, the mathematical progression from a complex to a simplified form, is contradicted by others who maintain that in practice it is difficult to know beforehand which terms to neglect or simplify.

The other suggestion is, therefore, to consider first a lesser number of degrees of freedom (thus reducing the number of original equations) and to try to find a solution with the help of as many simplifying assumptions as possible. The matter is complicated because each degree of freedom introduces, when mass inertia forces are considered, more unknown values than the number of mathematical equations, and the solution of these equations thus becomes indeterminate.

The best approach seems to be a combination of these two methods. We start by writing down all six equations of motion expressing the six degrees of freedom and, after a preliminary analysis, we neglect those degrees of freedom which from a practical aspect play an obviously less significant role. The remaining equations may be still further simplified by neglecting the terms which in the previous analysis of the steady state motion proved to be, in practice, small and hence insignificant. This method, although considerably reducing the number of unknowns, usually leaves too many for a definite solution. The only method then is to investigate the unknown values separately, or in conjunction with selected parameters, while temporarily keeping the remaining parameters constant. This permits the study of the inter-relationship between the parameters and leads to a better understanding of the complex relationship as a whole.

In this chapter a general scheme of such a method with a description of the 'state-of-the art', as far as the subject permits, is presented. Transient car motion is therefore presented in a schematic and dualistic way, giving on the one hand the relevant mathematical background, and on the other a qualitative or descriptive assessment of the phenomena.

We shall therefore consider the following six equations of car motion with respect to axes fixed relative to the ground.

* The car, when acted upon by a transient external force can, theoretically, move and accelerate in three-dimensional space, i.e. in three directions: longitudinal, transverse and vertical. It can also rotate angularly around these three axes. The theoretical possibility of either linear or angular motion is called the 'degree of freedom' of the motion. Thus, each of these six degrees of freedom may be expressed mathematically in the form of an equation of motion in the given direction. Such an equation describes the equilibrium either of forces or moments, including, in the transient case, inertia forces, i.e. accelerations.

(1) Transient motion in the OX (longitudinal) direction described by equating the longitudinal inertia force (which is the product of the car mass M and its acceleration $\mathrm{d}V/\mathrm{d}t = \dot{V}$) to the sum of all the forces acting in this direction, i.e.

(a) the tractive force T of the engine
(b) the mechanical (tyre rolling) drag D_m
(c) the aerodynamic drag force D_A.

Assuming that the mechanical drag is a 'weak function', i.e. only slightly dependent on speed, we may consider it constant. The aerodynamic drag force is transient and dependent on the variable airstream velocity V_A. Thus

$$M\dot{V} = T - D_m - D_A \tag{9.3}$$

(2) The transient motion in the OY (transverse) direction is described by equalising the transverse mass inertia force (the product of the car mass and its transverse accelerations $\mathrm{d}U/\mathrm{d}t = \dot{U}$ and the centripetal acceleration V^2/R) to the sum of the forces acting in this direction, i.e.

(a) the tyre side force Y_T
(b) the aerodynamic side force Y_A.

The tyre side force, or so called 'cornering force', is dependent on variables such as the side-slip angle β which is now a transient. Therefore the force Y_T is also dependent on time. Thus we have

$$M\dot{U} + M\frac{V^2}{R} = Y_T + Y_A \tag{9.4}$$

(3) The transient motion in the OZ (vertical) direction is described by equating the vertical inertia force (the product of the car mass M and the acceleration in the vertical direction $\mathrm{d}w/\mathrm{d}t = \dot{w}$) to all the other forces acting in this direction, neglecting suspension forces, i.e.

(a) the weight of the car W
(b) the time variable aerodynamic lift L_A.

Thus we have

$$M\dot{w} = L_A - W \tag{9.5}$$

(4) The transient yawing motion, i.e. the angular (rotation) motion about the vertical axis OZ (passing through the centre of gravity) is described by the inertia torque, i.e. the product of the moment of the car mass inertia about the OZ axis and the yawing angular acceleration $\mathrm{d}r/\mathrm{d}t = \dot{r} = \ddot{\beta}$, which is then equated to the sum of all other yawing moments, i.e.

(a) to the tyre moment N_T
(b) the yawing moments due to the coupling with the rolling and pitching moments $I_{zx}\ddot{\phi}$ and $I_{yz}\ddot{\alpha}$

(c) the moment due to the aerodynamic force N_A.

All these moments are variable with time. Thus we have

$$I_z\ddot{\beta} = I_{zx}\ddot{\phi} + I_{yz}\ddot{\alpha} + N_T + N_A \tag{9.6}$$

(5) The transient angular rolling motion which is the angular motion about the longitudinal axis OX is described by equating the torque due to the car mass inertia about the OX axis multiplied by the angular acceleration to

(a) the coupling moments due to yawing and pitching, $I_{xz}\ddot{\beta}$ and $I_{xy}\ddot{\alpha}$

(b) the rolling moment due to the centripetal acceleration of the suspended mass $M_s h V^2/R$

(c) the moment due to the acceleration in the transverse direction $M_s \dot{U} h$

(d) the rolling moment due to the aerodynamic side-force $Y_A d$.

All these moments are now variable in time. Thus we have:

$$I_x\ddot{\phi} = I_{xz}\ddot{\beta} + I_{xy}\ddot{\alpha} - M_s h(\dot{U} + V\dot{\beta}) + Y_A d \tag{9.7}$$

(6) The transient angular pitching motion, which is the angular motion about the transverse axis OY. It is described by equating the pitching inertia torque about the OY axis (passing through the C.G.) to the sum of all the pitching moments acting, i.e.

(a) the yawing and rolling coupling moments $I_{yz}\ddot{\beta}$ and $I_{xy}\ddot{\phi}$

(b) the longitudinal suspended mass inertia moment $M_s\dot{V}z_s$

(c) the aerodynamic pitching moment $L_A x_c + D_A z_c$.

All these terms are time variable. Thus we have

$$I_y\ddot{\alpha} = I_{yz}\ddot{\beta} + I_{xy}\ddot{\phi} + M_s\dot{V}z_s + L_A x_c + D_A z_c \tag{9.8}$$

where z_s is the height of the C.G. of the sprung mass above the position of the total mass C.G.

These six equations form the basis for calculating the response of a vehicle to a transient wind gust.

Introducing Equation 9.2 into the equation for the aerodynamic force, and then introducing into this expression the six equations of motion give six mathematical expressions which enable the time-motion behaviour to be analysed. The number of the time dependent variables involved in these equations of motion are greater than the number of equations, and also the equations are non-linear, i.e. are second degree differential equations and thus fairly difficult to operate mathematically. Consequently we have to make simplifying assumptions with a view to obtaining a more convenient mathematical form.

When discussing the wind velocity step we assumed that the velocity changes linearly. It is convenient to resolve the wind velocity which may blow at an angle ψ to the car trajectory into two components, one in the direction of motion and the other at right angles to it. We may assume that the longitudinal component does not greatly affect the car motion and can be neglected. By considering only the cross-wind we thus simplify our calculations. From Equation 9.3 and the definition of the aerodynamic force we develop the transient relative airstream velocity during a gust in the form

$$V_A = V_o + k_1 t \pm k_2 \cos \psi t^2 + k_3 \cos^2 \psi t^3,$$

where $k_{1,2,3}$ are constants dependent on the car characteristics and the slope of the velocity curve. For the cross-wind component we have $\cos \psi = 0$, and the expression is then simplified to

$$V_A = V_o + kt$$

which is consistent with the assumption of a linear velocity step. This time-dependent airstream velocity V_A may now be introduced into the equations of motion.

It is reasonable to assume that any acceleration in the vertical direction due to a horizontal wind gust would be negligible and thus Equation 9.5 can be neglected.

Further simplification may be obtained by neglecting the possibly small terms such as the pitching angular acceleration $\ddot{\alpha}$ or the rolling angular acceleration $\ddot{\phi}$. This helps to reduce the number of equations to five. These equations correlate the following values: the car velocity V, its side-slip angle β, the rate of yawing $r = \dot{\beta}$, the radius of the car trajectory R, the steering deflection angle δ, the side-slip velocity U, and the acceleration in the longitudinal and transverse directions.

The steering wheels deflection angle δ does not change the moment a gust of wind hits the car since, as mentioned earlier, there is a time-lag before the driver begins to respond. It is possible then to consider the angle δ as a constant. This allows us to reduce the problem to a single differential equation of the form

$$\dot{\beta}A + \beta B = Y_A C + D \tag{9.9}$$

The introduction of known values into Equation 9.9 allows the calculation of the corresponding aerodynamic side force Y_A. This force is calculated by introducing the parameters of the wind gust velocity which gives

$$Y_A = K_1 + K_2 t + K_3 t^2$$

where $K_{1,2,3}$ are constants dependent on the initial airstream velocity V_o, the aerodynamic coefficients, the car reference area, the air density and the rate of the air velocity increase during the gust. The time t is measured from the beginning to the end of the wind gust, i.e. the gust duration.

These equations allow us to develop a 'mathematical model' of the dynamic response characteristics for each vehicle under investigation. Such a model, with an appropriate computer programme, can be used to study the affect on dynamic stability not only of wind gusts but also of changes in the steering mechanism, tyres and the general dimensions and shape.

The characteristic response is shown in Fig. 9.4 which gives the yaw angular velocity in non-dimensional units and the yaw angle and rolling angle responses to the aerodynamic impulse. All these are plotted against a time scale. The rate of yaw and roll are shown as being much greater for cars characterised by a high aerodynamic lift coefficient.

The actual side-slip angle variations are much dependent on other characteristic dimensions such as the wheelbase. Thus the diagram shows the response as the accumulation of different factors. Racing cars and long vehicles show much greater side-slip angle response than ordinary saloons. Cars with larger lift coefficients also show a comparatively large roll, yaw and side-slip response.

The variation in response between different vehicle types is considerable. Fig. 9.4 shows the response to external aerodynamic impulses and what is of considerable interest to note is that the different types do not show any degree of automatic damping. Only the roll response seems to indicate a certain automatic damping tendency. Thus the yawing response is practically undamped. Cars once deflected from their original attitude remain in their new attitude and only the driver's response in changing the steering wheel deflection angle can correct this.

The fact that road vehicles behave reasonably well under transient motion conditions, at first slowed down the theoretical investigation of their dynamic characteristics. It was found that the problems in designing dynamically stable vehicles could be solved very well by experimental investigation. Experimental work in the field of dynamic stability is now being pursued with increasing vigour. Thus, whether new or old models, all road vehicles are now tested practically for their response to external forces, or their response to steering impulses, and the behaviour is measured in terms of angular yawing rates, or side-slip angle changes. When the vehicle so tested shows unpleasant mannerisms different modifications are tried until the fault is reduced or corrected. These tests are accumulating data about the behaviour of road vehicles under a variety of external impulsive conditions which are then compared with advances in the theoretical work. This will in time perfect the 'mathematical model' with which we work [86], and cars will then be much safer to drive, in good or bad weather.

It is, therefore, of interest to survey the present state of experimental research in car dynamic stability.

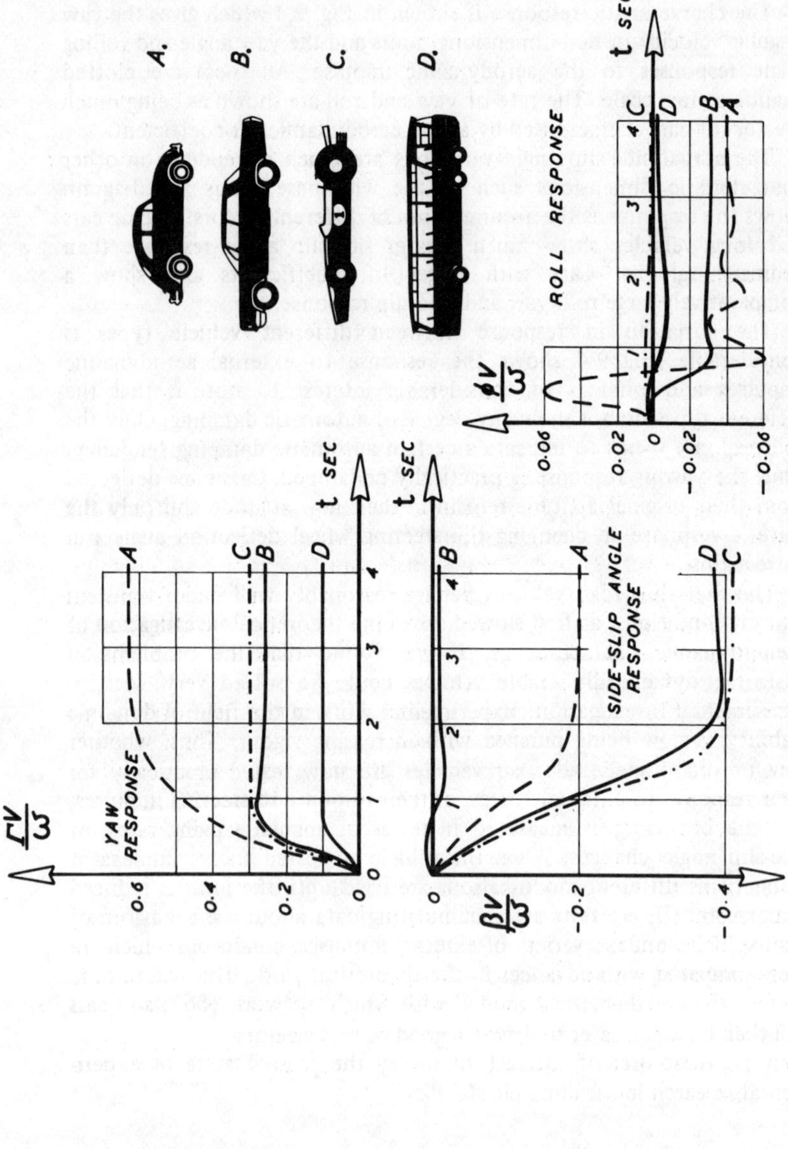

Fig. 9.4. The yaw, roll and side-slip responses of typical vehicles

9.5 PRESENT STATE OF EXPERIMENTAL RESEARCH INTO DYNAMIC STABILITY

The theoretical and experimental research into car dynamic stability has a comparatively short history. The part which includes the impulsive external aerodynamic force has a particularly short history. The theoretical analysis of this problem is complex and therefore difficult. This makes the experimental approach very attractive and sometimes necessary. Experimental investigation requires methods of simulating an impulsive aerodynamic force. Some work on this subject has been published in various papers, although, all in all, comparatively little. These papers were published between 1953 and 1959 and all were concerned with wind gust effects on cars in motion [73,74,76,86].

One of the earliest efforts to simulate the impulsive aerodynamic force used the car weight as the external force. The side force can be considered as a weight component when the car tilts sideways. This can be achieved by building a road track with varying camber, Fig. 9.5(a). The car is run at speed on such a ramp and is then subjected to a linear side-force step, the size of which depends on the camber angle of the road. The test track on which such experiments were carried out should not have a camber greater than $+4°$. The gravitational side force is then $W \sin 4°$ and produces a step from zero to $0.07\,W$ which, for a vehicle weighing 1000 lbf, represents an increase of 70 lbf.

This method of simulating an external impulse does not satisfy the requirements of correct simulation. The side-force component operates at the centre of gravity of the vehicle and not at the aerodynamic centre of pressure–these two points are often at a distance from each other. Further the resultant weight which presses the car wheels on the ground decreases as the road camber increases and as the tyre side-slip characteristics are very sensitive to the changes in this load it introduces a considerable error. In addition we have, at the very moment the step in the side force is applied, an elastic reaction of the suspension which by its consequent recoil affects the behaviour of the car in a different way to the transient aerodynamic force.

It seems therefore that the results obtained by these experiments are misleading to a considerable degree. The typical results of measurement of transient values of the yaw angular velocity r, the side-slip angle β, and the roll angle ϕ measured by the use of this method and plotted on a time scale are shown in Fig. 9.5(b). Comparing these results with those obtained by computing the theoretical equations indicates a degree of similarity between them. The responses obtained, and especially the roll angular velocities and angles, are however greater than the theoretical values and of reversed sign. This might be due to the suspension recoiling.

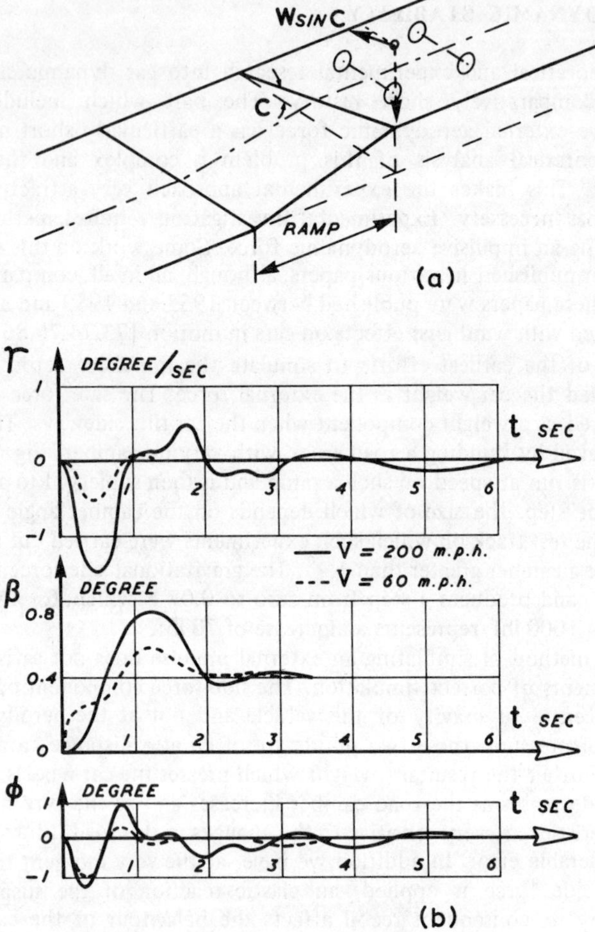

Fig. 9.5. (a) Variable camber road as a method of simulation of an external transient force; (b) yaw and roll responses measured by the variable camber road method

Basic theoretical and experimental work on dynamic stability has also been undertaken at the Research Laboratories of the General Motors Corporation in U.S.A., and the results published in 1963 [74]. In these investigations the external transient aerodynamic force was simulated by the concentrated thrust force of a hydrogen-peroxide rocket. This rocket motor was fixed with its nozzle axis at right angles to the car plane of symmetry and at the estimated point of the aerodynamic centre of pressure. This method of simulation had the advantage that the effect of the position of the C.P. could be investigated. The shortcomings of this method are many. First, the aerodynamic force is the result of the pressure distribution and thus is strictly related to the airflow pattern around the vehicle. In the dynamic wind gust condition, as discussed previously, there is a time gap between the change of the flow pattern and the change of pressure distribution. This phenomenon has a considerable influence on the variable force effect and simulation which does not reproduce these conditions may lead to different results. Secondly, the weight of the rocket motor and its installation changes the actual distribution of weight and thus the car inertia characteristics. The moments of inertia around all three axes of the vehicle play an important role in the equations of motion as we can see from the relevant equations.

Assuming that the experiments produced the effects of a side-force Y_A on the car, we have still to remember that these results are unrelated to the shape of the vehicle. Nevertheless, this method of simulating the external aerodynamic force allows a certain check on the theoretical results. It makes possible also the study of the effect of the position of the centre of pressure relative to the centre of gravity. It has proved the sensitivity of the yaw angular velocity responses, and consequent changes in the trajectory, to the positioning of the C.P. In general it is found that when the C.P. is downstream of the centre of gravity the actual impulse of the aerodynamic force is more effective in increasing the initial rate of yaw but it introduces an element of damping by quickly decreasing the actual aerodynamic side force. When the C.G. and C.P. coincide, the aerodynamic side force will not affect the rate of change of the yaw angle (the aerodynamic yaw moment about the C.G. is then zero) but it will increase the tendency to push the car sideways, i.e. to add to its side-slip velocity.

The artificial wind gust method

The only correct method of simulating aerodynamic transient forces is to create the conditions of an external wind gust which can be controlled and measured. Then we can subject full sized cars to any given wind gust condition at will. The known size and intensity of the artificial wind gust allows the correct instrumentation to be used both on the car and on the road.

In practice it is very difficult to create wind gusts in a wind tunnel and virtually impossible to incorporate the necessary instrumentation into a scale model. A further problem would be the difficulty in simulating an elastic suspension and the mass inertia characteristics of a full-sized vehicle. We are therefore compelled to use full-sized vehicles in our experimental techniques. This makes it particularly difficult to investigate transient aerodynamic effects on a car prototype prior to its manufacture.

The experimental investigation of dynamic stability has to be carried out on existing car types, and this accumulated data and experience can then be used for future design. Thus, experimental investigations are limited to checking the theory rather than direct application in the design.

The first attempt to use an artificial wind gust simulator was made in Stuttgart in about 1950. The device used for this purpose consisted of an aircraft jet engine mounted with its axis at right angles to the test track. The exhaust gases were blown across the track and the vehicles under test in passing through this gas jet were thus subjected to a known 'velocity step'. The vehicles were equipped with instruments measuring the three directional and angular accelerations. The single jet engine produced only a limited gust size–too small for a car to be properly affected by it–and the so-called 'gust volume' was not sufficient for the not too sensitive recording. In other words the response time of the recording instruments was too long for the short duration aerodynamic impulse instruments used. The ratio of the length of track affected by the jet of exhaust gases to the car length was of order unity. This resulted in only a fraction of the vehicle being affected for a short time before running almost directly into the negative velocity step at the other end of the jet. The total effect was to simulate buffeting rather than wind gust conditions.

The ratio of the car length to the size of the gust, i.e. the gust volume should be of the order of 5 to 30 if we try to simulate the natural phenomenon. Such a gust volume permits the instruments to measure all the transient changes involved, and thus a correct analysis of the single velocity step effect can be made.

The solution of this problem was found in the concept of a multiple jet engine installation. This system produces a jet exhaust covering longer track distances, providing the blending of each engine exhaust is correct and the cross-wind velocity distribution is more or less controlled. This factor was considered when designing a slightly less expensive installation which consists of a single large jet engine which is positioned close to a test track at MIRA and blows directly across it. Two alternative widths of gust can be set up. The first is a narrow gust of 20 ft (6 m) nominal width and 50 m.p.h. (80 km/hr) mean speed which has been used to test vehicles against an American experimental safety vehicle specification. In these tests a maximum path deviation is set for a vehicle driven through the gust with steering held fixed. Whilst the specified path deviation provides an absolute standard, most of the tests made have been comparative, between different vehicles or vehicle configurations.

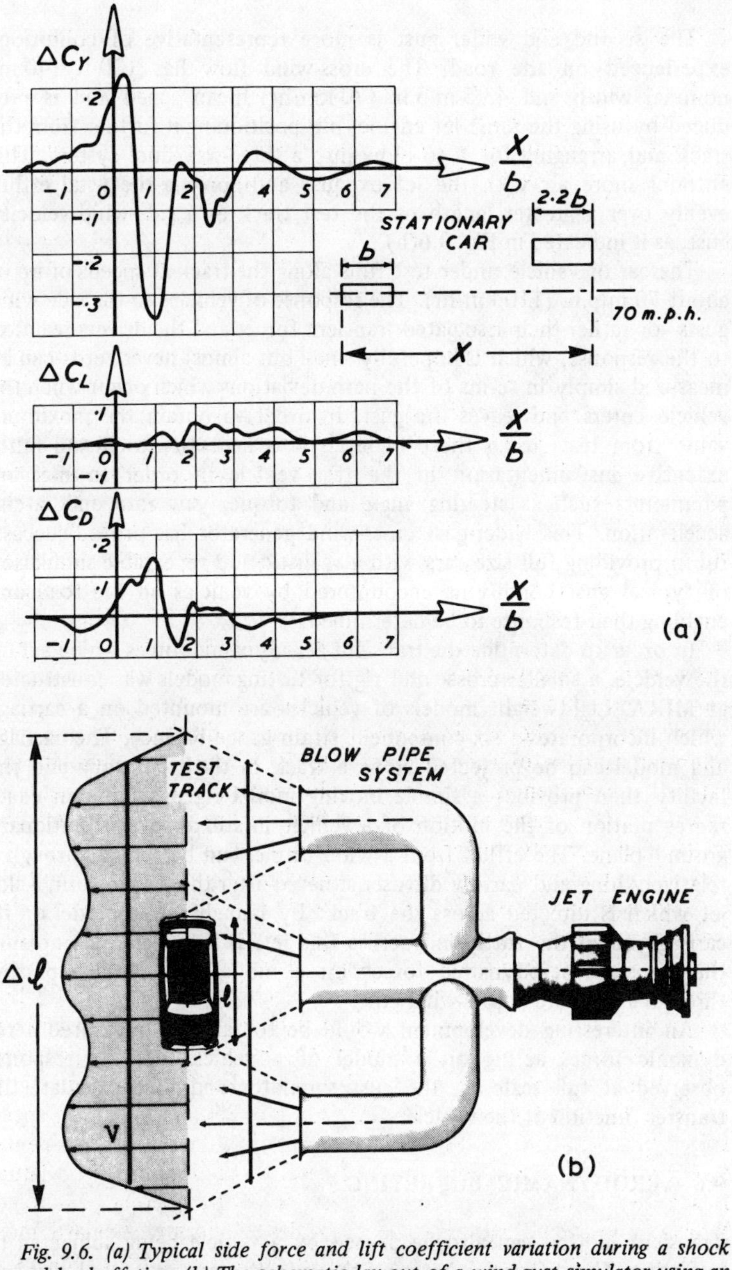

Fig. 9.6. (a) Typical side force and lift coefficient variation during a shock caused by buffeting. (b) The schematic lay-out of a wind-gust simulator using an aeronautical jet engine

The second and wider gust is more representative of conditions experienced on the road. The cross-wind flow has 130 ft (40 m) nominal width and 43.5 m.p.h. (70 km/hr) mean speed and is produced by using the same jet engine, but positioning it further from the track and arranging for it to blow into a three-way duct system. This entrains more air with the jet exhaust and spreads the total efflux evenly over a greater length of the test track as a side-wind velocity gust, as is indicated in Fig. 9.6(b).

The car or vehicle under test runs along the track at speeds of up to about 70 m.p.h. (110 km/hr). The response of vehicles to the side wind gusts—or rather their associated transient forces and the drivers reaction to the response, which is hopefully small but almost never zero—can be measured simply in terms of the path deviations which occur when the vehicle enters and leaves the gust. In order to obtain the maximum value from tests using these facilities it is necessary to install fairly extensive instrumentation in the test vehicle in order to measure parameters such as steering angle and torque, yaw rate and lateral acceleration. This wider-gust cross-wind generator has proved successful in providing full size cars with a realistic and repeatable simulation of typical gust conditions encountered by vehicles on the road and enabling their response to be determined.

In order to determine the transient aerodynamic forces which act on the vehicle, a smaller cross-wind rig for testing models was constructed at MIRA. Lightweight models of vehicles are mounted on a carriage which incorporates a six-component strain gauge balance. The carriage and model can be projected along a track in the laboratory and the facility then provides a simple moving model rig providing an exact representation of the motion of a vehicle in still air over a stationary ground plane. The efflux from a wind tunnel can be spread through a relatively long and narrow diffuser situated after the fan, to form a slot jet which is directed across the track. By launching the model on its carriage down the track and across this jet, it is possible to measure the transient aerodynamic forces on a model of a vehicle passing through a simulated cross-wind gust.

An interesting development would be to take the measured aerodynamic forces acting on a model of a vehicle and the response observed at full scale on the gust simulator and then calculate the transfer function of the vehicle.

9.6 AERODYNAMIC BUFFETING

Transient motion can also be caused by a vehicle being subjected to buffeting. The difference between the velocity step in a wind gust and buffeting lies basically in their duration. When the car is affected by a 'buffet', it means that it is subjected to a single and sudden periodic change of pressure, and consequently a change in the aerodynamic force. In other words it is like an initial velocity step followed immediately by another of opposite sign, the beginning and the end of a wind gust linked together. There are many occasions on the roads where conditions which cause buffeting are met. The most common is the pressure wave created by, and travelling with, a vehicle when passing another car. The interference between two such flow patterns when vehicles are passing each other at speed causes buffeting to occur. Turbulent wakes trailing behind vehicles come also into this category. It is known, in practice, that such pressure waves can be quite intensive and that they can cause momentarily a quite rapid change of pressure distribution which produces a transient aerodynamic force. There have been cases of cars falling off their jacks (when changing wheels) due to other vehicles passing nearby. British Rail investigated a case some time ago of carriage windows which were shattered by violent air buffeting after two trains each travelling at about 90 m.p.h. passed each other. Passage at speed can cause cars to swerve and such conditions are often encountered on motorways.

The problem of aerodynamic buffeting was investigated experimentally among others, by the Ford Motor Company [29] and by the Von Karman Institute in Belgium. The experimental method employed was to move two scale models relative to each other. This was done by mounting one of the models rigidly and fixing the other to a large pendulum which swung along a curved path passing the fixed model at a controlled distance. The aerodynamic forces and their coefficients due to aerodynamic buffeting of the fixed model are shown in Fig. 9.6(a). This diagram shows that the increase of the side-force coefficient during the buffet is far greater than the changes of lift and drag coefficients which are much less significant. These exemplary results were obtained for bluff bodies representing railway locomotives, but they are of the same character as those for the more streamlined saloon car. The maximum pressure induced by one moving vehicle on the other in the 'buffet' periodic pressure change is dependent on their relative speed. The buffeting intensity and duration of the pressure changes depend also on the shape of the vehicle. It has been established experimentally that fully streamlined car shapes produce more concentrated pressure fields and greater pressure variations than blunt bodies which cause a flow disturbance in much larger volumes of the surrounding air, and hence smaller pressure differences. Thus aerodynamic buffeting is more intense in amplitude and frequency for highly streamlined cars.

The effect of aerodynamic buffeting can be compared to the effect of an intense but short wind gust. The ratio of the size of the buffet zone to the car size is approximately one–much smaller than in the case of a wind gust. The problems of buffeting require more experimental research since although it is of less importance than the wind gust effect, nevertheless, it is of a similar character and can influence road vehicle safety.

9.7 CONCLUSION

The problems of the transient state have been comparatively well elaborated from the point of view of vehicle steering response. On the aerodynamic side extensive work is required to establish not only typical values of aerodynamic derivatives, but also the relation between particular values of a given derivative and the body shape. Thus the effects of styling on the intensity and duration of transient state motion should be studied systematically.

Before such work is completed typical values of the derivatives should be found and the importance of aerodynamic effects in the transient state more accurately evaluated. These effects are considerable as anyone would confirm having driven on a windy day on a motorway full of large-sized lorries. We have to obtain numerical criteria and clear numerical relationships between characteristics.

It seems therefore that a considerable part of the experimental work on road vehicle aerodynamics will in time pass from wind tunnels and tests on scale models to investigations carried out on full-sized vehicles on the test track. This requires the development of efficient instrumentation and further improvement in measuring techniques.

REFERENCES

1. Klemperer, W. 'Investigations of the aerodynamic drag of Automobiles' *Zeitshrift fur Flugtechnik,* **13,** (1922).
2. 'Lehrstuhl und Forshungsinstitut fur Kraftsfahrtwesen und Fahrzeugmotoren' *Prospectus, Technishe Hochshule,* Stuttgart (1967).
3. Carr, G. W. 'Aerodynamic lift characteristics of cars', Paper No. 1, *Meeting Auto. Div. Inst. Mech. Engrs.* May (1973).
4. Barth, R. 'Effect of shape and airflow about an automobile on drag, lift and directional control', *Verein Deutches Ingenieure,* **98,** No. 22, (1956).
5. Kamm, W. and Schmid, C. *Automobile Testing,* Springer, Berlin (1938).
6. Barth, R. 'Stromform und Gebrauchswagen', *Motor Rundshau* Jan. (1959).
7. Barth, R. 'Effect of unsymmetrical wind incidence on aerodynamic forces acting on vehicle models and similar bodies', *Soc. Automot. Engs.,* U.S.A. Yearbook (1966).
8. Neesen, F. *Gestaltung und Wirtschaftlichkeit der Land, Wasser und Luftfahrzeugen',* Jena Verlag (1940).
9. Barth, R. 'The aerodynamics of car body shapes', *Engineers Digest,* **17,** No. 10 (1956).
10. Schmid, C. 'Luftwidrstand am Kraftfahrzeugen', Versuche am Fehrzeug, und model' (Air resistance on cars and models), Deutche Kraftfahrtforshung: Verein Deutches Ingenieure Verlag, No. 1., (1938).
11. Bekker, M. G. *Theory of Land Locomotion,* The University of Michigan Press (1956).
12. Scibor-Rylski, A. J. 'Experimental investigation of the negative aerodynamic lift wings used on racing cars', *Advances in Road Vehicle Aerodynamics, Br. Hydromech. Ass.,* Cranfield, (1973).
13. Harrison, N. 'The effectiveness of forward negative lift devices on a racing car', Unpublished, The City University (1973).
14. Tietjens, O. G. 'Economy of streamlining the automobile'. *Soc. Automot. Engs., Trans.,* **30,** No. 3, (1932).
15. Carr, G. W. 'The aerodynamics of basic shapes for road vehicles', *Mot. Ind. Res. Ass.,* Rep. No. 4 (1970).
16. Deutch, C. *Dynamique des Vehicules Routieres,* Organisme National de Securite Routiere, (1971).
17. Greenroyd, J. S. 'A study of the airflow patterns over surfaces at an angle to each other', Unpublished, The City University (1970).
18. Scorer, R. S. *Natural Aerodynamics,* Pergamon Press (1958).
19. Hoerner, S. F. *Fluid Dynamic Drag,* Published by Author, New York (1951).
20. Streeter, V. L. *Handbook of Fluid Dynamics,* McGraw Hill, New York (1961).
21. Prandtl, L. and Tietjens, O. G. *Applied Hydro and Aeromechanics,* Dover, New York (1935).
22. Prandtl, L. and Tietjens, O. G. *Fundamentals of Hydro and Aeromechanics,* Dover, New York (1935).
23. Stapleford, W. R. and Carr, G. W. 'Aerodynamic characteristics of exposed rotating wheels', *Mot. Ind. Res. Ass.,* Rep. No. 2 (1970).

24. Morelli, A. 'Aerodynamic actions on an automobile wheel' Paper 105, *Proc. 1st Symposium on Road Vehicle Aerodynamics,* The City University, London (1969).
25. Loghton, C. J. 'Rotating wheel experiment', Unpublished, The City University, London (1969).
26. Thwaites, B. *Incompressible Aerodynamics,* Clarendon Press, Oxford (1960).
27. Bassett, A. B. *A treatise on Hydrodynamics,* Dover, New York (1961).
28. Milne-Thopson, L. M. *Theoretical Hydrodynamics,* McMillan, London (1949).
29. Scibor-Rylski, A. J. *Proc. 1st Symposium on Road Vehicle Aerodynamics,* The City University, London (1969).
30. Koenig-Fachsenfeld, F. R. *Aerodynamik des Kraftfahrzeuges,* Frankfurt am Main, Motor-Rundshau Verlag (1951).
31. Bagley, J. 'The pressure distribution on two-dimensional wings near the ground', *Aeronaut. Res. Coun.* Rep. No. 3238, (1960).
32. Sounders, G. H. Aerodynamic characteristics of wings in ground proximity', *Can. Aeronaut. Space Inst. J.,* June (1965).
33. Stephens, H. S. *Advances in Road Vehicle Aerodynamics, Br. Hydromech. Ass.,* Cranfield (1973).
34. Tustin, R. C. and Carr, G. W. 'Aerodynamics of basic shapes for small saloon cars', *Mot. Ind. Res. Ass.,* Rep. No. 10 (1963).
35. White, R. G. S. 'A rating method for assessing vehicle aerodynamic drag coefficients', *Mot. Ind. Res. Ass.,* Rep. No. 9 (1967).
36. Cornish, J. J. 'Some aerodynamic characteristics of land vehicles', *Am. Inst. Aeronaut. & Astronaut. Symp.,* Los Angeles (1968).
37. Stratton, D. G., Stringer, R. E. and Taylor, S. R. G. 'Engine cooling system design and development', *Proc. Auto. Div. Inst. Mech. Engrs.,* **180,** Pt. 2A, No. 8 (1965).
38. Kays, W. M. and London, A. L. *Compact Heat Exchangers,* 2nd ed. McGraw Hill, New York (1964).
39. Gehres, E. 'An analysis of engine cooling in modern passenger cars'. *Soc. Automot. Engrs.,* Paper No. 660C March (1963).
40. Fishenden, M. and Saunders, O. A. *'An Introduction to Heat Transfer',* Oxford Univ. Press (1950).
41. Taylor, S. R. G. 'A study of vehicle engine cooling', Advanced School of Automobile Engineering, Thesis (1963).
42. Latwell, P. V. 'Investigation into the internal air flow through the cooling system of an automobile', Unpublished, The City University, London (1967).
43. Berry, A. J. 'An experimental investigation of ground effect using a moving surface simulation', Thesis, Imperial College, London (1968).
44. Romberg, G. F., Chianese, F. and Lajoie, R. G. 'Aerodynamics of race cars in drafting and passing situation', Paper No. 710213, *Soc. Automot. Engrs. Congress,* Detroit (1971).
45. Scibor-Rylski, A. J. 'Aerodynamic characteristics of the wedge shaped racing car', *Automot. Des. Eng.,* Sept. (1970).
46. Pershing, B. 'Dragster aerodynamic streamlining versus weight', *Am. Inst. Aeronaut. & Astronaut. Symp.,* Los Angeles (1968).
47. Newcomb, T. P. and Spurr, R. T. *Braking of Road Vehicles,* Chapmann and Hall (1966).
48. Scibor-Rylski, A. J. 'Body of minimum drag in the proximity of ground', Unpublished, The City University, London (1971).
49. Kamm, W. and Goudert, K. 'The streamlined car', *Motor Life Magazine,* New York, Oct. (1954).

50. Klemin, A. 'A belt method of representing the ground', *Aero Sci. J.*, **1**, No. 4, Oct. (1934).
51. Klemin, A. 'Moving ground developed for tunnel testing of car models', *Automot. Industry,* **71**, No. 5, (1934).
52. Kelly, K. B. J., Kyropoulos, P. and Tanner, W. F. 'Automobile aerodynamics', *Soc. Automot. Engrs.*, Rep. No. 148 B,C,D,F, March (1960).
53. Sykes, D. M. 'The effect of low flow rate gas ejection and ground proximity on afterbody pressure distribution', *1st Symp. Road Vehicle Aerodynamics,* The City University, London (1969).
54. Barth, R. 'Aerodynamic characteristics of windscreen wipers', *Auto. Technishe Zeitung,* No. 66/11, 323 (1964).
55. Fosberry, R. A. C. and White, R. G. S. 'The aerodynamics of road vehicles, a comparison of cars and their models in wind tunnels', *Mot. Ind. Res. Ass.*, Rep. No. 9 (1958).
56. Carr, G. W. and Rose, M. J. 'Correlation of full scale wind tunnel and road measurements of aerodynamic drag', *Mot. Ind. Res. Ass.*, Rep. No. 5 (1964).
57. White, R. G. S. 'An experimental survey of vehicle aerodynamic characteristics', *Mot. Ind. Res. Ass.*, Rep. No. 1 (1962) and No. 4 (1964).
58. Shapiro, A. H. *Shape and Flow, the Fluid Dynamics of Drag,* Doubleday, New York (1960).
59. Heltemes, C. J. 'Auto aerodynamics', *Car Life Magazine,* U.S.A. Sept.–Oct. (1961).
60. Kamm, W. 'Directional stability of motor vehicles', Tech. Memo. No. 103, Stevens Institute of Technology, Hoboken, New Jersey (1953).
61. Kamm, W. 'Vehicle characteristics affecting directional stability of motor vehicles', *Auto. Technishe Zeitung,* **60**, No. 6 (1958).
62. Segel, L. 'Theoretical prediction and experimental substantiation of the response of the automobile to steering control', *Soc. Automot. Engrs.*, 310 (1956).
63. Whitcomb, D. W. and Milliken, W. F. 'Design implications of a general theory of automobile stability and control', *Inst. Mech. Engrs. Year Book,* 367-391 (1956).
64. Baier, W. H. 'Vehicle steering fundamentals', *Trans. Soc. Automot. Engrs.*, **69**, 256-7 (1961).
65. Hales, F. D. 'Stability problems of road vehicles', *1st Symp. Road Vehicle Aerodynamics,* The City University, London (1969).
66. Ellis, J. R. 'Understeer and oversteer', *Auto. Engr,* 178-182, May (1963).
67. Russell, J. B. and Scibor-Rylski, A. J. 'Aerodynamic effects on the lateral control and stability of motor vehicles', *1st Symp. Road Vehicle Aerodynamics,* The City University, London (1969).
68. Russell, J. B. 'The effect of aerodynamic forces and moments on the lateral control and stability of motor vehicles', The City University, Memo. No. Ae1, Aug. (1968).
69. Flynn, H. and Kyropoulos, P. 'Truck aerodynamics', *Trans. Soc. Automot. Engrs.*, **70**, 297-308 (1962).
70. Whitcomb, D. W. 'The influence of aerodynamic forces on the critical speed of automobiles', *Cornell Aero. Lab. Inc.*, Rep. No. YM2262-F-1. May (1967).
71. Barth, R. 'Fahrzeugheizungen', *Auto. Technishe Zeitung,* Heft 11, Nov. (1962).
72. Bowman, W. D. 'Present status of automobile aerodynamics in automobile engineering and development', *Am. Inst. Aeronaut. & Astronaut.*, Symposium, Los Angeles (1968).

73. Milliken, W. F. and Whitcomb, D. W. 'General introduction to a programme of dynamic research', *Soc. Automot. Engrs.*, Yearbook, 287-309, (1956).
74. Bundorf, R. T., Pollock, D. E. and Hardin, M. C. 'Vehicle handling response to aerodynamic inputs', *Trans. Soc. Automot. Engrs.*, **72**, 7-73 (1964).
75. Segel, L. 'The variable stability automobile, concept and design', *Trans. Soc. Automot. Engrs.*, **74**, No. 650-658 (1966).
76. Lay, W. E. and Letts, P. W. 'Wind effects on car stability', *Trans. Soc. Automot. Engrs.*, **61**, (1953).
77. Evans, R. D. 'Properties of tyres affecting riding, steering and handling', *Trans. Soc. Automot. Engrs.*, **43**, Feb. (1935).
78. Olley, M. 'Road manners of the modern car', *Proc. Instit. Automot. Engng.*, **51**, 147, (1946/7).
79. Shilling, R. 'Directional control of automobiles', *J. Ind. Math. Soc.*, **4**, (1954).
80. Kelly, K. B. and Holcombe, H. J. 'Aerodynamics for body engineers', *Soc. Automot. Engrs.*, Paper No. 649 (1963).
81. Beauvais, F. N., Garelis, C. and Jacovoni, D. H. 'An improved analog for vehicle stability analysis', *Soc. Automot. Engrs.*, Meeting, Detroit (1961).
82. Zbrozek, J. K. 'Some effects of atmospheric turbulence on aircraft', *Weather*, **13**, No. 7 (1958).
83. Wax, M. P. 'An experimental study of wind structure', *Electl. Res. Ass.*, Tech. Rep. C/T 114 (1956).
84. Pankhurst, R. C. and Holder, D. W. 'Wind tunnel technique', Pittman (1948).
85. Durand, W. F. *Aerodynamic Theory*, Vol. 3 and Vol. 4, Springer, Berlin (1935).
86. Slibar, A. and Paslay, P. R. 'Behaviour of vehicles subjected to wind gusts', *Ingenieur Archives*, No. 28 (1959).
87. White, R. G. S. 'A rating method for assessing vehicle aerodynamic side force and yawing moment coefficients', *Mot. Ind. Res. Ass.*, Rep. No. 1 (1970).
88. Cogotti, A. 'Aerodynamic characteristics of car wheels', *Int. J. of Vehicle Design*, Technological Advances in Vehicle Design Series, SP3, Impact of Aerodynamics on Vehicle Design, 173-196 (1983).
89. Muto, S. 'The aerodynamic drag coefficient of a passenger car and methods of reducing it', *Int. J. of Vehicle Design*, Technological Advances in Vehicle Design Series, SP3, Impact of Aerodynamics on Vehicle Design, 57-69 (1983).
90. Howell, J. 'Wake properties of a saloon car', *Proc. Koll. Industrie Aerodynamik*, Aachen, Pt. 3: Aerodynamik van Strassenfahrzeugen, 85-95 (1974).
91. Bearman, P. W., Davis, J. P. and Harvey, J. K. 'Measurement of the structure of wakes', *Int. J. of Vehicle Design*, Technological Advances in Vehicle Design Series, SP3, Impact of Aerodynamics on Vehicle Design, 493-499 (1983).
92. Hucho, W. H. 'The aerodynamic drag of cars–current understanding, unresolved problems and future prospects', *Aerodynamic Drag Mechanisms of Bluff Bodies and Road Vehicles*, 7-44, Plenum Press, New York (1978).
93. Carr, G. W. 'Potential for aerodynamic drag reduction in car design', *Int. J. of Vehicle Design*, Technological Advances in Vehicle Design Series, SP3, Impact of Aerodynamics on Vehicle Design, 44-56 (1983).
94. Morel, T. 'The effect of base slant on the flow pattern and drag of three-dimensional bodies with blunt ends', *Aerodynamic Drag Mechanisms of Bluff Bodies and Road Vehicles*, 191-217, 226, Plenum Press, New York (1978).
95. Carr, G. W. 'Reduction of cooling system aerodynamic power losses in road vehicles', Pt. 3 of Vehicle Fuel Economy: Effects of Aerodynamics and Gearing, Department of Industry, London (1976).

96. Stapleford, W. R. 'Aerodynamic improvements to the body and cooling system of a typical small saloon car', *Proc. 4th Coll. Industrial Aerodynamics, Aachen, Road Vehicle Aerodynamics*, 121-132 (1980).
97. Roussillon, G. 'Aerodynamic optimisation of the Peugeot V.E.R.A.', *Int. J. of Vehicle Design*, Technological Advances in Vehicle Design Series, SP3, Impact of Aerodynamics on Vehicle Design, 114-131 (1983).
98. Paish, M. G. and Stapleford, W. R. 'A study to improve the aerodynamics of vehicle cooling systems (second report)', MIRA Report 1968/4 (1968).
99. Wright, P. G. 'The influence of aerodynamics on the design of Formula One racing cars', *Int. J. of Vehicle Design*, Technological Advances in Vehicle Design Series, SP3, Impact of Aerodynamics on Vehicle Design, 154-172 (1983).
100. Choulet, R., Crosnier, S. and Schafer, G. 'FISA-évaluation de la relation règlement-performance pour la Formule Un', *SERA CD*, Paris (1982).
101. Bolster, J. 'Theory becomes practice', *Autosport*, 3rd Feb., 48, 49 (1983).
102. Carr, G. W. 'Aerodynamics as a means to vehicle fuel economy', Paper C210, *Function versus Appearance in Vehicle Design*, I.Mech.E., Conference Publications 1978-7, London (1978).
103. Hucho, W.-H. 'Designing cars for low drag-state of the art and future potential', *Int. J. of Vehicle Design*, Technological Advances in Vehicle Design Series, SP3, Impact of Aerodynamics on Vehicle Design, 1-8 (1983).
104. Bahnsen, U. 'Airflow management in future design', *Int. J. of Vehicle Design*, **3**, No. 1, 40-47 (1982).

BIBLIOGRAPHY

Andrade, J. 'Wind tunnel tests and body design', *J. Soc. Automot. Engrs.*, 29 (1931).

Bodger, W. K. and Jones, C. M. 'Aerodynamic wind throb in passenger cars', *Soc. Automot. Engrs.*, Automobile week (1964).

Brown, L. H. 'Economics of streamlining in heavy transportation', *Soc. Automot. Engrs.*, Meeting, May (1931).

Bundorf, R. T. 'The use of the variable stability vehicle in handling research', *Trans. Soc. Automot. Engrs.*, 74, No. 650-659 (1966).

Carr, G. W., Rose, M. J. and Smith, N. P. 'Some aerodynamic aspects of safety in road vehicles', *Proc. Inst. Mech. Engrs.*, 187, No. 30 (1973).

Carr, G. W. 'Aerodynamic effects of underbody details on a typical car model', *Mot. Ind. Res. Ass.*, Rep. No. 7 (1965).

Carr, G. W. 'Aerodynamic effects of modifications to a typical car model', *Mot. Ind. Res. Ass.*, Rep. No. 4 (1963).

Carr, G. W. 'The MIRA quarter-scale wind tunnel', *Mot. Ind. Res. Ass.*, Rep. No. 11 (1961).

Cato, W. C. and Meek, J. F. 'Improvement of fuel economy by minor changes in aerodynamic shape', *Soc. Automot. Engrs.*, Detroit meeting, Feb. (1961).

Flanagan, D. L. and Hefner, O. V. 'Surface molding, new tool for the engineer', *Astronaut. and Aeronaut.*, Yearbook (1967).

Fosberry, R. A. C. 'The aerodynamics of road vehicles' (A survey of published literature), *Mot. Ind. Res. Ass.*, Rep. No. 1 (1958).

Fosberry, R. A. C. and Holubecki, Z. 'The aerodynamics of road vehicles', (Road measurement of drag comparison with tunnel measurements), *Mot. Ind. Res. Ass.*, Rep. No. 10 (1958).

Fosberry, R. A. C. and White, R. G. S. 'A 250 H.P. Chassis Dynamometer', *Mot. Ind. Res. Ass.*, Rep. No. 7 (1963).

Fosberry, R. A. C., White, R. G. S. and Carr, G. W. 'A British wind tunnel installation and its application', *Soc. Automot. Engrs.*, Paper No. 948C, *Int. Automot. Engrs.*, Congress, Detroit (1965).

Fosberry, R. A. C. and White, R. G. S. 'The MIRA full-scale wind tunnel', *Mot. Ind. Res. Ass.*, Rep. No. 8 (1961).

Gebelein, H., Hahn, H. and Schlick, G. 'Increased stability and directional control contributes to greater vehicle safety', *Deutche Kraftfahrtforshung und Strassenverkehr*, Technik, No. 122, Verein Deutches Ingenieure, Verlag, Dusseldorf (1959).

Glauert, H. *The Elements of Aerofoil and Airscrew Theory*, 2nd Ed., Cambridge Univ. Press (1948).

Goetz, H. 'The influence of wind tunnel tests on body design ventilation and surface deposits on sedans and sports cars', *Soc. Automot. Engrs.*, Paper No. 710212 (1964).

Gough, V. E. 'Cornering characteristics of tyres', *Auto. Engr.*, Apr. (1954).

Hales, F. D., Barter, N. F. and Skinner, M. J. 'The handling and stability of motor vehicles', Pts. I, II, IV, V, VI, VII, *Mot. Ind. Res. Ass.*, Rep. No. 1 (1965), Rep. No. 8 (1965), Rep. No. 6 (1967), Rep. No. 1 (1968), Rep. No. 7 (1968), Rep. No. 10 (1970).

Hales, F. D. 'Vehicle handling qualities', *Proc. Inst. Mech. Engrs.*, London (1969/70).

Hales, F. D. 'Automobile stability', T.M.150 and 151, Davidson Lab., Hoboken (1958).

Hillier, P. 'A study of the effects of cabin position on the forces and moments acting on a motor car', Unpublished, The City Univ., London, June (1968).

Jacobs, R. 'Experimental investigation on racing car model', Unpublished, The City Univ., London (1968).

Jeans, Sir James. *An Introduction to the Kinetic Theory of Gases,* Cambridge Univ. Press (1940).

Kofmann, J. L. and Aron, K. 'Fan position' and 'Cooling air flow', *Auto. Engr.*, Jan. (1956).

Koffman, J. L. 'Automobile fans', *Auto. Engr.*, Jan. (1955).

Koffman, J. L. 'Liquid cooling', *Auto. Engr.*, Feb. (1954).

Korff, W. H. 'The aerodynamic design of the Goldenrod to increase stability, traction and speed', *Am. Inst. Aeronaut. and Astronaut.*, Symposium, Los Angeles, Apr. (1968).

Korff, W. H. 'The body engineers role in automotive aerodynamics', *Soc. Automot. Engrs.*, Paper No. 649b (1963).

Kraemer, W. E. 'Flugelprofile im kritishen Reynoldszahl', *Forshung aus dem Gebiete Ingenieurvesehs,* **27**, No. 2 (1967).

Larrabee, E. E. and Hawks, R. J. 'The calculated effect of cross-wind gradients on the disturbance of automotive vehicles', *Am. Inst. Aeronaut. & Astronaut.*, Symposium, Los Angeles, Apr. (1968).

Lay, W. E. 'Is 50 miles per gallon possible with correct streamlining?', *J. Soc. Automot. Engrs.*, **32** (1933).

Marva, R. S. 'A study of the negative lift device for racing cars', Unpublished, The City Univ., London (1969).

McLean, R. F. and Shilling, R. 'Behind the Firebird car's new look', *J. Soc. Automot. Engrs.*, **62**, Aug. (1954).

Mitschke, M. 'Directional control and road stability in four wheeled motor vehicles', *Mot. Ind. Res. Ass.*, Trans. 14/61 (1961).

Moller, E. 'Drag measurements on the Volkswagen delivery truck', *Auto. Technishe Zeitung.*, **53**, No. 6, (1951).

Moller, E. 'Wind tunnel measurements on automobiles in a cross-wind', *Auto. Technishe Zeitung.*, **53**, No. 4, (1951).

Nagler, L. H. and Razak, K. 'Full scale automobile aerodynamics', Pt. 1, *Motor,* March 8, (1950) and Pt. 2 March 15, (1950).

Paish, M. G. and Stapleford, W. R. 'A rational approach to the aerodynamics of engine cooling design', *Proc. Inst. Mech. Engrs., Auto. Div.*, **183**, Pt. 2A, No. 3 (1968).

Paish, M. G. and Stapleford, W. R. 'Ram induced airflow through ducted and non-ducted radiator matrices, *Mot. Ind. Res. Ass.*, Rep. No. 15, (1966).

Paish, M. G. and Stapleford, W. R. 'Ducted fan and Radiator design' and 'Application of ducted fan cooling system to a medium size family saloon car', *Mot. Ind. Res. Ass.*, Rep. No. 4, (1968).

Papin, P. and Sachs, H. K. *Proc. 2nd Int. Conf. Vehicle Mechanics,* Paris, Sept. (1971).

Pope, A. *Wind Tunnel Testing,* Wiley, New York (1954).

Raviolo, V. G. 'Motor car of the future', *Engineer,* **217**, No. 5594, 656-7 (1963).

Riegels, F. W. *Aerofoil Sections,* Butterworth (1961).

Riekert, P. and Gauss, F. 'Bestimmung des Luftwiderstands von Kraftfahrtzeugen durch Auslaufsversuche in vergleigh mit Modellmessungen', *Auto. Technishe Zeitung,* **54**, No. 10 (1952).

Romani, I. 'Les Essais aerodynamiques des voitures automobiles, maquettes et vehicles reels', *Union Technique de L'Automobile et du Cycle,* Paper No. 11 (1950).

Sachs, H. K. *Proc. 1st Int. Conf. Vehicle Mechanics,* Wayne State Univ., July (1968).

Schlichting, H. 'Aerodynamic problems of motor cars', *Advisory Group Aerospace Res. & Dev.,* Rep. No. 307 (1960).

Scibor-Rylski, A. J. 'If cars had wings', *New Scientist,* June, 12 (1969).

Scibor-Rylski, A. J. 'Negative lift devices on racing cars', Paper No. 19, *Symp. Road Vehicle Aerodynamics,* The City Univ., London (1969).

Scibor-Rylski, A. J. 'Tour of European Centres of Industrial Aerodynamics', Report to the Goldsmith Co., Unpublished (1969).

Scibor-Rylski, A. J. 'The aerodynamics of racing cars', Internal publication, The City Univ., London (1968).

Scibor-Rylski, A. J. 'Economy of automobile aerodynamics', *Autocar,* 34-36, May, 18 (1967).

Scibor-Rylski, A. J. and Clark, R. P. 'The experimental investigation of a method to measure pressure distribution over a car model', Internal Memo, The City Univ., London, May (1967).

Spear, P. and Rudd, A. C. 'Some aspects of motor racing research', Engineering Development Facility, Rubbery Owen (1966).

Strunk, W. *The Elements of Style,* Macmillan, New York (1959).

Teagle, P. R. 'The effect of the internal flow through the body on flow characteristics and drag of a simplified car model in a wind tunnel in the presence of a rigid board simulating the ground', Unpublished, The City Univ., London, July (1972).

Townsend, A. A. *Structure of turbulent shear flow,* Cambridge Univ. Press (1956).

White, J. F. 'New techniques for full scale testing', *Soc. Automot. Engrs.,* National Automobile Week, Detroit, March (1960).

White, R. G. S. and Paish, M. G. 'A study of vehicle aerodynamic lift', *Mot. Ind. Res. Ass.,* Rep. No. 16 (1966).

SYMBOLS

a	distance between the centre of gravity and front axle
b	distance between centre of gravity and rear axle
d	distance between aerodynamic centre of pressure and roll axis
f	rolling drag coefficient
g	gravitational acceleration
h	distance between centre of gravity and roll axis
$h.p.$	engine brake-power
l	wheel-base distance
p	pressure
$q = \frac{\rho V_A^2}{2}$	dynamic pressure of the airstream
r	yaw angular velocity
$\dot{r} = \frac{\mathrm{d}r}{\mathrm{d}t} = \frac{\mathrm{d}^2\beta}{\mathrm{d}t^2}$	acceleration in yawing angular motion
t	time
$t_{F,R}$	wheel track distance of front and rear axles
w	atmospheric wind speed
$x, y, z,$	three-dimensional space coordinates
x_b	horizontal distance between C.G. and C.P.
z_c	vertical distance between C.G. and C.P.
A_T	self-aligning wheel torque
$A_{TW} = \frac{\partial A_T}{\partial W}$	self-aligning torque derivative with respect to change of weight
$A.R.$	aspect ratio (i.e. the ratio of width to length)
C_D	aerodynamic drag coefficient
C_{Di}	coefficient of induced drag
C_L	aerodynamic lift coefficient
C_m	pitching moment coefficient
$C_{m\alpha} = \frac{\partial C_m}{\partial \alpha}$	derivative of pitching moment coefficient with respect to pitch angle
C_N	yawing moment coefficient
$C_{N\beta} = \frac{\partial C_N}{\partial \beta}$	derivative of yaw moment coefficient with respect to side-slip angle

$C_p = \dfrac{p - p_o}{q}$	pressure coefficient
C_q	rolling moment coefficient
C_Y	aerodynamic side-force coefficient
$C_{Y\beta} = \dfrac{\partial C_Y}{\partial \beta}$	derivative of side-force coefficient with respect to side-slip angle
C.G.	centre of gravity
C.P.	point of application of aerodynamic force (centre of pressure)
H	total pressure of the airstream
$I_{z,y,x}$	moments of car inertia about *oz*, *oy*, and *ox* axes
I_{zx} I_{xy} I_{yz}	polar moments of car inertia
K	static margin (distance expressed as a fraction of wheel base between C.G. and line of action of resultant tyre and aerodynamic forces)
L_A	aerodynamic lift force
M	pitch moment (about C.G.)
M_o	total mass of car
M_s	sprung mass of car
N_A	yaw moment due to aerodynamic forces
N_T	yaw moment due to tyre frictional reaction
P	rolling drag of tyres
$P_W = \dfrac{\partial P}{\partial W}$	tyre rolling drag derivative with respect to change of weight
Q	rolling moment coefficient (Q_A when it is of aerodynamic origin)
R	radius of car turning trajectory
$R_e = \dfrac{\mu V_A l}{\rho}$	Reynolds number (l is a characteristic arbitrary dimension)
S	reference sectional area of car
T	tractive force of driving wheels
U	sideways velocity (side-slip)
$\dot{U} = \dfrac{dU}{dt}$	acceleration in sideways direction
V	vehicle speed with respect to ground
$\dot{V} = \dfrac{dV}{dt}$	acceleration of car in longitudinal direction
V_A	airstream velocity relative to car
W	weight of car
Y_A	aerodynamic side-force
Y_T	frictional (cornering) force of tyres

$Y_{A\beta} = \dfrac{\partial Y_A}{\partial \beta}$	aerodynamic side-force derivative with respect to side-slip angle
$Y_{A\phi} = \dfrac{\partial Y_A}{\partial \phi}$	aerodynamic side-force derivative with respect to roll angle
$Y_{A\delta} = \dfrac{\partial Y_A}{\partial \delta}$	aerodynamic side-force derivative with respect to deflection angle of steering wheels
$Y_{Tr} = \dfrac{\partial Y_T}{\partial r}$	tyre side-force derivative with respect to yaw angular velocity
$Y_{TW} = \dfrac{\partial Y_T}{\partial W}$	tyre side-force derivative with respect to change of weight
$Y_{T\beta} = \dfrac{\partial Y_T}{\partial \beta}$	tyre side-force derivative with respect to yaw angle
$Y_{T\phi} = \dfrac{\partial Y_T}{\partial \phi}$	tyre side-force derivative with respect to roll angle
$Y_{T\delta} = \dfrac{\partial Y_T}{\partial \delta}$	tyre side-force derivative with respect to deflection angle of steering wheels
α	angle of pitch
β	side-slip (or yaw) angle of car
δ	angle of deflection of front steering wheels
$\delta_{w,g}$	the thickness of the viscous boundary layer
μ	the viscosity coefficient of air or frictional coefficient of tyre on road
ψ	angle between relative airstream and car axis
θ	reference angle between direction of wind and radius of turning
ζ	yawing, or side-slip angle of single wheel (i.e. angle between plane of rotation and direction of wheel motion)
ρ	air mass density
ϕ	angle of roll
η	coefficient of efficiency of engine

INDEX